For God and Glory

With bible read and final prayer written, Nelson ponders his mortality before the Battle of Trafalgar, 21 October 1805.

For God and Glory

LORD NELSON
AND
HIS WAY OF WAR

Joel Hayward

NAVAL INSTITUTE PRESS
Annapolis, Maryland

Naval Institute Press
291 Wood Road
Annapolis, MD 21402

Library of Congress Cataloging-in-Publication Data
Hayward, Joel S. A.
For God and glory : Lord Nelson and his way of war / Joel Hayward.
p. cm.
Includes bibliographical references and index.
ISBN 1-59114-351-9 (acid-free paper)
1. Nelson, Horatio Nelson, Viscount, 1758–1805—Military leadership.
2. Naval art and science—Great Britain—History—18th century. 3.
Great Britain—History, Naval—18th century. 4. Naval
tactics—History—19th century. 5. Admirals—Great Britain—Biography.
I. Title.
DA87.1.N4 H37 2003
940.2'745'092—dc21

2002015688

Printed in the United States of America on acid-free paper ⊗
10 09 08 07 06 05 04 9 8 7 6 5 4 3 2

Contents

Preface

I live with my family in a pleasant house on the corner of two quiet streets. There is nothing unusual about that, of course. Other people share this type of situation. Yet the two streets forming our corner bear the names of warriors from a war that ended almost two hundred years ago—warriors successful in their own right, but most famous for their close relationships with one of history's greatest heroes. We live on the corner of Collingwood and Hardy Streets; but as few of our neighbors probably realize, Collingwood and Hardy were seamen who played important roles in the life of Vice Adm. Horatio Lord Viscount Nelson (1758–1805). Cuthbert Collingwood was Nelson's lifelong friend and his second in command at the climactic Battle of Trafalgar in October 1805. Thomas Hardy, another of his long-term friends, was the splendid flag captain who ably served Nelson in several campaigns and who gave him the famous kiss on the forehead as he lay dying.

One might imagine from our street names that we live somewhere in England, Nelson's beloved homeland. England has hundreds of streets, squares, malls, and parks named after the half-blind, one-armed admiral and his "band of brothers," as he lovingly called his subordinate captains. But we aren't English; we live in Palmerston North in distant New Zealand. This small and peaceful nation in the southwest Pacific became known to Nelson's compatriots through the voyages of Captain Cook in the 1760s and 1770s, but was not settled by them until two generations after Nelson's tragic death at Trafalgar in 1805.

Yet New Zealand, which effectively began its national life in 1840 when English settlers entered into a formal treaty with the indigenous Maori, soon established a number of towns and cities and gave some the names of the mother country's favorite sons. Cook Strait, for example, the stretch of water that separates New Zealand's main islands, was obviously named in honor of the famous navigator, as was our nation's highest mountain. The cities that lie on each side of Cook Strait were named after Britain's greatest warriors: Nelson and Wellington.

The streets that form our corner were not named in the 1840s or 1850s, when nostalgia for Napoleonic battles and heroes remained high. They were named in the 1950s, a full 150 years after the untimely death of the great seaman. Such is the magic and mystique surrounding Nelson that even Collingwood and Hardy, two of his close colleagues, were receiving honors a century and a half later as well.

As for Nelson himself, I cannot think of another military leader whose life is commemorated so richly two centuries after his death. Every 21 October the Royal Navy and most of its Commonwealth counterparts celebrate Trafalgar Night, a formal Mess Dinner in honor of Nelson and his victory and death at the Battle of Trafalgar on 21 October 1805. Even some American naval squadrons and fleets celebrate Trafalgar Night, treating Nelson with the same reverence as they give their own heroes, John Paul Jones and Stephen Decatur.

It is hard to find a naval base throughout the world that does not somewhere—either on the wall of commanders' offices or in the wardrooms or libraries—have Nelson's distinctive image displayed. I always enjoy seeing his framed portrait hanging above the main doors leading into the wardroom of Her Majesty's New Zealand Ship (HMNZS) Devonport, New Zealand's main naval base. I also like seeing the paintings of his most famous engagements and battles—the Gulf of Genoa, the Hyères, Cadiz, Cape St. Vincent, the Nile, Copenhagen, and of course Trafalgar—lining the wardroom walls.

Military and naval historians in many nations have chronicled Nelson's life and dissected aspects of his character and career in hundreds of biographies, monographs, and journal articles. Indeed, he has received more scholarly and literary attention than any other military or naval figure, except possibly for Napoleon Bonaparte. The National Maritime Museum in Greenwich, London, has become virtually a shrine to Nelson's memory, and his flagship, HMS *Victory,* magnificently restored in Portsmouth dockyard, remains a popular tourist attraction. Two British societies exist to promote the memory of his glory: the Nelson Society (founded in 1981) and the 1805 Club (founded in 1990). Both have members around the world, hold lectures and meetings, and publish journals packed with interesting and often important articles on Nelson and the Georgian navy. They also have Web sites on the Internet that are attracting ever-growing numbers of hits. And Nelsoniana, that seemingly endless outpouring of Nelson ceramics, toby jugs, busts, statuettes, paperweights, postcards, and other collectibles, easily outsells the commemoratives of every other military and naval leader with the possible exception, once again, of Napoleon.

It is easy to see why Horatio Nelson became a legend in his own lifetime, a hero to rival Napoleon, and an enduring symbol (indeed, almost the personification) of sea power's influence and potential. He lived in an age when a successful admiral might gain one victory in fleet or squadron action during his lifetime. Yet he secured a victory for Sir John Jervis at Cape St. Vincent (1797) and then gained three more of his own: at the Battle of the Nile (1798), the Battle of Copenhagen (1801), and the Battle of Trafalgar (1805). His flair for tactics and his mastery of battle were dazzling. The scale and significance of his victories were also unparalleled. The taking of nineteen French and Spanish ships off Cape Trafalgar gave him more enemy ships than any of his predecessors had captured or destroyed in all their battles put together, and, more importantly, it secured for Britain mastery of the sea for over a century. It is no wonder that he is naval history's grandest figure and the most well-known and revered warrior in Britain's pantheon of heroes.

But one could legitimately ask whether Nelson is *relevant* today, almost two hundred years after his death and ninety after Britain surrendered its naval mastery to other nations, particularly the United States. Today's world is very different from Nelson's. Ever-evolving technologies and ideas, often themselves emerging from wars and conflicts, have changed the way we perceive international relations, regional security, national identities, and even our own communities. These technologies and ideas have certainly changed the concepts, methods, and machinery of warfighting. Naval vessels, for instance, barely resemble the lumbering, wind-reliant, constantly leaking, and rotting warships of Nelson's time. Driven by nuclear energy or gas turbine power plants, naval vessels today have almost unbelievable range, speed, and maneuverability. Thanks to satellites and shipboard tracking and sensor equipment, these vessels have fantastic battle space awareness. They also carry an array of nuclear and conventional weapons of frightening destructive potential. The doctrines guiding their employment are different from and vastly more sophisticated than those that underpinned the fleet activities of Nelson's era. Is there any merit, then, in studying the naval tactics and fighting styles of that bygone age?

Our values and ethics are also markedly different. Since around the time of the Vietnam War, most Western nations have become a little less willing to commit their armed forces to conflicts unless the cause is (or can be *sold* as) primarily humanitarian and not merely national and political. Public expectations of warfare have also changed. Whereas in Nelson's era death and privation were accepted as ever present features of war, now the public wants ironclad guarantees of minimal casualties, no collateral damage, and the quick return of combatants. The growing Western sensitivity to casualties has been manifest on many occasions in the quarter-century since Vietnam. The clearest case is perhaps the dramatic drop in American public support for the United Nations peace enforcement mission in Somalia, which followed the October 1993 killing of eighteen American soldiers in Mogadishu. President Clinton and the United States Congress responded promptly by withdrawing the entire American force. So what value is there in studying the life of Nelson, a commander famous for his love of battle and for the level of destruction he inflicted?

Even the way historians record the deeds of armed forces has changed. The new military history that sprang up in the 1960s as a subbranch of social history focuses far less on what senior officers and commanders did, and more on the everyday experiences, motivations, views, and feelings of ordinary soldiers, sailors, and airmen. This refocusing was, and still is, necessary. For quite some time, the genre of military and naval history has not been held in high regard by many professional historians, who believe it lacks rigor, contains far too much anecdotal and emotive narrative, needlessly creates heroes and villains, and, worst of all, glorifies warfare.

It is certainly true that much military and naval history is poor, a problem apparently stemming from the inadequacies of the authors themselves. Many are buffs with little or no historical training or former soldiers and seamen who cannot maintain any critical distance from past events. These amateurs focus most of their attention on particular combat units and the sharp end of military operations with little regard for context or issues of truth, objectivity, and bias.

Many also write uncritical and eulogistic books on the lives of *great captains,* men of purported genius like Alexander, Caesar, Napoleon, Nelson, Grant, Lee, Rommel, and Patton. The writers become so enamored with their subjects that they greatly exaggerate their subjects' importance, presenting them not so much as products of their national, social, and cultural milieus, but more as creators or at least shapers of them. Treating the terms *biography, hagiography* (recording the lives of saints), and *history* as virtual synonyms, these writers produce naïve and simplistic historical analyses of the lives and times of their heroes. Psychological and behavioral consistency and plausibility do not seem to matter. Glorious attributes or events are covered at length, while ignoble or mediocre ones are passed over quickly or even misrepresented.

Given that most studies of great captains are therefore no longer considered valuable and useful by many in the academic and literary communities, and even by some in areas related to defense, another legitimate question might be raised. Is there merit in my adding another title specifically on Lord Nelson to the large number already written?

I hope to demonstrate that the questions I have asked can all be answered in the affirmative. Nelson and his remarkable way of war *are* relevant to today's world and worthy of further analysis, particularly by contemporary warriors, but also by anyone interested in the role of human personality in situations of chaos and conflict.

Since around the time of the Persian Gulf War of 1991, new weapons, information technologies, and tactics have purportedly been transforming warfare to enable battles to be fought over the horizon with little face-to-face contact between belligerents. Military professionals commonly describe this transformation as the Revolution in Military Affairs, or RMA. The more ardent and technocentric proponents of this RMA claim that the fundamental nature of warfare is changing (see Schneider and Grinter, eds., *Battlefield of the Future*). Digitization and remotely operated weapons, they say, have already made redundant the chaotic, frightening, and violent close combat that has always characterized warfare. The norm will now be less chaotic and more manageable stand-off battles (hitting the enemy from positions of safety) and long-range precision strikes.

Other more cautious commentators, including me, accept that the means of prosecuting warfare have changed dramatically, so that a joint or combined-arms battle today would be incomprehensible to, say, one of Frederick the Great's soldiers. Yet we also believe that warfare's fundamental *nature* has not changed at all. Rather than being in a military revolution, it seems clear that we are merely in a period of rapid change in the technologies available with which to do the same old jobs: intimidating, coercing, punishing, taking, crushing; in short, threatening or inflicting bloodshed.

Military history is replete with similar periods of rapid change in the technologies and corresponding techniques and tactics of warfare. The introduction of gunpowder to Western armies and navies, for instance, improved their killing power and gave them significant advantages over those still using handheld, thrown, or projected weapons. During the nineteenth century, industrialization—manifest mainly in railroads, the telegraph, the steam engine, rifled weapons, and ironclad ships—also improved human killing power and changed the techniques and tactics of battle. More recently, the development of internal combustion engines and powered flight ushered in blitzkrieg, carrier aviation, and strategic bombing. All these advances, and their resulting changes to the techniques and tactics of battle, increased human killing power even further.

Yet throughout all these periods of rapid change, conventional close combat encounters remained central to battle, the face of which has scarcely altered. Warfare has always been a human activity. Weapons change, but *they* do not fight each other. The humans using them do. More efficient and effective tools—many of them exploiting the ever-improving power of the microchip—may now make humans able to wage war with infinitely greater awareness of battle space, but these weapon-using *individuals* will still be fighting and wounding and killing. War has not changed. It is still a remarkably human activity—regretfully—waged by persons motivated by subjective factors of great complexity (a fine starting point on this topic may be found in Evans and Ryan, eds., *Human Face of Warfare*).

There is tremendous merit, therefore, in learning what our forebears have done, even in long-ago eras like Nelson's, as well as studying the guiding principles of war that have shaped battle and conflict throughout the centuries. The study of history does not provide fast-food solutions—instantly accessible, prepackaged, and easy to digest—to the challenges of armed conflict (as suggested in Harper and Pratten, *Still the Same*). The study of the past is hard work, but immensely rewarding. Servicemen and women may get only one chance to do what they have been trained and prepared for: to fight, survive, and win. So they must be fully informed about both the nature of the combat in which they may find themselves and the options for dealing effectively with various circumstances that arise. Failure to be informed may prove fatal. Past experiences, if studied thoughtfully in their proper contexts and in an interdisciplinary framework, provide safe and priceless opportunities for exploring the complex and subjective factors that influence combatants and determine their performance.

There is much value in understanding war in terms of the everyday experiences, views, and feelings of ordinary soldiers, sailors, and airmen. After all, they do the fighting, they risk their lives, they suffer wounds, and they do the dying. Yet, they do so within formal hierarchies that place decision makers over them. Those decision makers—their commanders—manage the ordinary troops' lives and shape their battles. Their commanders order the advances, attacks, withdrawals, and retreats. They assume responsibility for their troops' safety, welfare, and morale. This responsibility is a heavy burden for the commanders and is seldom treated casually.

One should not shy away, then, from trying to understand the characters and careers of senior commanders, even the great captains like Nelson, who not only bestrode their own eras but also made advancements in warfighting concepts and the practice of what we call "the profession of arms." The roles of commanders in the wars of their eras might be considered a poor distinction, given that they are the ones who planned, ordered, and directed the battles that caused (or *prevented*) so much destruction and misery. Yet these commanders warrant scholarly assessment precisely because they did so.

Sir Basil Liddell Hart, one of Britain's most distinguished military commentators, once wrote that if military history is to be of practical value, "it should be a study of the psychological reaction of the commanders, with merely a background of events to throw their thoughts, impressions and decisions into clear relief" (*Thoughts on War*, 219). I think this is going a little far. Those who study military history should analyze what *all* participants felt, thought, and did. They should look at soldiers in trenches and foxholes, and at seamen and airmen in their own equally unpleasant situations. They should also look at the generals, admirals, and air marshals, who usually enjoy more comfort but who have accepted the weighty burden of responsibility for their subordinates' lives. Accounts of the characters and careers of senior commanders are as essential to gaining a full understanding of warfare's many complexities as are the splendid "war from below" accounts now popular.

Yet, the sort of conventional biography just mentioned is not always the best explanatory tool. If getting inside a commander's head is wanted, a more useful and illuminating means of attaining that goal is to adopt a thematic approach, allowing the subject's motivating passions, influential ideas, and patterns of behavior to be explored individually and in depth, rather than merely in passing as they emerge in traditional biographical studies.

My goal in this book is to do precisely that: to analyze Lord Nelson's fascinating and influential life, with particular focus on his style of warfare and the passions, ideas, beliefs, and behavioral patterns that created and shaped that style, via a thematic approach. Perhaps I have given some themes undue importance. I confess to having reflected long and hard before settling on those I chose to explain. But I accept that the scholarly process is by its very nature subjective, and this book represents my own best efforts to understand Lord Nelson and his way of war.

Having read virtually every major biography and most journal articles written on Nelson, as well as having pored over the collections of his and his colleagues' seemingly innumerable diary entries, letters, and dispatches, I am acutely aware that Nelson has been splendidly served by his biographers. I warmly praise Carola Oman, Oliver Warner, Tom Pocock, Christopher Hibbert, Colin White, and others for their outstanding contributions to scholarship. They have assiduously tracked down long-forgotten and hitherto-unused source materials and explicated all relevant contextual issues. With some exceptions, these biographers have avoided the hagiography that plagued their nineteenth-century forerunners. Their accounts have certainly enriched my knowledge and sparked my imagination. I thank them.

Following in their footsteps has been a daunting task. But I respectfully suggest that my book has something to offer that differs significantly from their excellent studies. It manages this through its thematic approach, which has allowed me to break free of chronology's sometimes constraining clutches to explore in greater-than-usual depth and coherence the key aspects of Nelson's warfighting style. This book, a work of warfighting analysis more than of traditional history, also asks several new questions about that style and its supporting ideas.

I work as a professional military analyst and historian and research, write, and teach on the tactics and operational art of land warfare, sea power, and joint warfare. I humbly suggest that my training and experience allow me to view Nelson's warfighting from vantage points not recognized by, or accessible to, many of his leading biographers. Their experience and interests tend to give them a great grasp of naval tactics (upon which I have often relied) but not necessarily of other themes and topics in warfare. I have thus asked and attempted to answer a number of questions, which include the following:

What did the admiral think of Britain's and his own enemies? How consistent were his responses to them? Did they change over time? What were his motives for staying at war—and with such unrelenting commitment—long after physical disabilities made this difficult?

Did his deep Christian faith significantly influence his way of war and his treatment of superiors, subordinates, and enemies? What impact did his abiding religiosity have on his warfighting and leadership styles, his thoughts on destruction and death (ever present in war, of course), and his commitment to king, kin, and country?

How did he view and practice his interrelated responsibilities as commander, leader, and manager? Were his instinctive, unrivaled talents for inspiring and motivating other people in combat situations matched by an ability to exercise authority prudently and manage time, men, and matériel effectively?

What were the doctrinal ideas, integral elements, and influence of his idiosyncratic warfighting style? To what degree does that warfighting style, certainly effective in his era, resemble that now considered *best practice?* Is there anything that modern warriors can learn from Nelson's ideas, experiments, experiences, successes, and failures?

How successful were his attempts to fight battles on land? To what degree did he understand that the basic principles of naval warfare were (and still are) very different from those employed by soldiers? Did he understand that weather, terrain, logistics, fortifications, and other factors significantly influenced the methods adopted on land?

How warm and congenial was his relationship with the army? Did he understand and respect the army's contributions to campaigns, and was he willing and able to work harmoniously with military counterparts? Was his marvelous flair for battle at sea matched by similar instincts and abilities on land?

Was Nelson able to build strong relations and establish and maintain goodwill in multinational coalition contexts? Did the admiral, who was exceedingly patriotic and often critical of the abilities and efforts of Britain's allies, manage to see other national viewpoints, adopt conciliatory policies, set common priorities, and establish multinational strategies?

My interpretations, reinterpretations, and conclusions may not be accepted by all readers, but I will be delighted if these speculations prompt others interested in Nelson to seek their own answers.

Some readers may notice that I use certain modern professional military and naval phrases to describe or explain what Nelson was doing two hundred years ago. At first glance this may seem anachronistic and understandably give traditional historians and biographers some discomfort. Yet it is not wrong to use our greater comprehension and range of specialized terminology to explain Nelson's way of war in a manner that today's naval planners, practitioners, and enthusiasts best understand. It is no more anachronistic than a modern scientist using today's language and knowledge to explain what Thomas Edison was doing over a century ago. We now have terms, phrases, and defined concepts to illuminate the principles that proved difficult back then to conceptualize or explain.

Finally, my term *warfighting style* needs brief explanation. It better fits what I'm trying to wrestle with than do terms such as doctrine, tactics, and leadership. Although these are important topics in my analysis, they define observable group activities. As terms they pertain to what all members of a unit or formation do as a group in coordinated fashion based on either explicit or implicit principles—as well as on inspiration, training, and instructions.

By warfighting style I mean something far more individualistic. I mean a way of doing things. It includes—but is far more personal and intangible than—the doctrine and tactics that shape group activities. Warfighting style is an individual commander's unique interpretation of doctrine and his selective use (or rejection) of established tactics. It is also his art of leadership, which is more than merely what he is (his characteristics or traits) and what he does (his observable behaviors). His *style* is also the way his traits allow him, often subconsciously, to use certain behaviors in different contexts to achieve various desired results.

A commander's style—this strange practical outworking of traits, ideas, ego, and ethos—is very important to a fighting unit. Following a change in leaders, a fleet or army may suddenly become either less or more disciplined, contented, audacious, aggressive, and ultimately effective simply because the new commander introduced a different style and way of getting things done. For instance, when Robert E. Lee took command of the Army of Northern Virginia following the wounding of Gen. Joseph E. Johnston in mid-1862, Lee's personality and style had a profound effect. His doctrine and tactics were not significantly different than those of his predecessor, and the strategic framework was something he could not immediately change. Yet Lee possessed an intangible quality, an ethos, a set of attitudes, and a way of getting things done that quickly transformed a competent army into a marvelous instrument of war with a cohesion, sense of purpose, and esprit de corps that was second to none.

This book seeks to describe and explain Lord Nelson's highly unique warfighting style. Doctrine, tactics, and operational art form part of the analysis, as do the admiral's command and leadership abilities and his attitudes and beliefs. But the focus is on how all these things evolved and combined during successive combat experiences to form the man whose infectious ethos—like Lee's several generations later—spread through his entire force as if carried on invisible currents. Even Lee appears colorless when compared to Nelson, whose creative genius, excitable and intense personality, deliberately dramatic visage, and fervor for all things martial did more than inspire loyalty and courage (which are no mean feats); Nelson dazzled, confounded, electrified, and inflamed the hearts and minds of his men—creating mythology that reached almost cult status even within his own lifetime.

Acknowledgments

The Creator—worthy of all blessings—deserves my greatest thanks. His love and grace have helped my family cope during troubled times in recent years. And what a family I have! My sweet-natured wife and best friend, Kathy, and my beautiful daughters have loved, encouraged, and assisted me as I researched and wrote this book. I return their love in abundance and offer them my devotion and sincere thanks.

I also thank my boss, mentor, and friend, Maj. Gen. Piers Reid, NZA (Ret.), CBE, Litt.D. This accomplished soldier and defense analyst cast his expert gaze over the entire manuscript and made many suggestions for improvement. General Reid's unparalleled knowledge of military, naval, joint history, strategy, and operational art has enriched all my scholarly projects over the last few years, including this book.

Dr. Glyn Harper (Lt. Col., NZA [Ret.]) also read my manuscript and recommended ways of clarifying certain points and making the entire book more user friendly. I am very grateful to him.

Colin White, deputy director and head of museum services at the Royal Navy Museum, Portsmouth, offered advice on some sections of the manuscript. I dared not ignore that advice; Mr. White knows more about Nelson than almost anyone alive. Thank you, Colin.

My thanks to Mr. Randolf Slaff of the Line of Battle Museum, Annapolis, Maryland, for making copies of various engravings available to me for this book. Paul Lumsden created my clear and attractive maps, for which I am grateful, and Sandra Scott, my ever helpful secretary-administrator, did a great job of word processing.

I'm very grateful to Naomi Grady for her excellent copyediting, which has improved my text significantly.

Last, but certainly not least, I thank Paul Wilderson, executive editor of the Naval Institute Press, for seeing merit in my project and accepting that my thematic approach, while unorthodox, allows key aspects of Nelson's warfighting style and experiences to be analyzed in more depth than traditional biographical methods usually allow. Thanks for your faith in me, Paul. I am honored to have my humble effort published by your press, the recognized world leader in naval studies.

For God and Glory

Introduction

THE PRIMARY focus of this book is to analyze Horatio Nelson's warfighting style, one that was unique to him and made him a great leader and England's most brilliant admiral.

The characteristic elements of that style are assessed in the following chapters, which treat separate but closely interrelated themes. Each chapter is an essay and focuses on the single theme indicated in its title, as that theme has been abstracted from the admiral's actions, behavior, and documented comments. But even though his warfighting style must be described in words, the style itself is not an abstraction. It is revealed in the events of a commander's life, in his behavior, his relationships, his decisions. Thus, it is important, before considering the themes of Nelson's warfighting style, to review briefly the events of the life itself in order to place those themes in context. For readers already familiar with the story of Nelson's life, the following thumbnail sketch of the major events in which Nelson participated will enable them to review the historical sequence. For those not familiar with Admiral Nelson's life, the overview of his professional endeavors will provide a basic background, enabling readers to understand his style of warfighting as it was expressed in the major events, battles, and campaigns in which he participated.

Nelson was born on 29 September 1758 at Burnham Thorpe, in Norfolk, England, the son of an Anglican clergyman and his wife. In 1771, three years following his mother's untimely death, he went to sea as a twelve-year-old midshipman. At the age of eighteen, after having served on a merchant ship (1771), an arctic explorer (1773), and several warships in the West and East Indies (1773–77), he gained promotion to lieutenant. Because of his exceptional ability, energy, and commitment, he received command of a sloop and then a frigate, and promotion to captain, all before he turned twenty-one. In 1780 the young sea officer (now made post captain) took part in an arduous military expedition up the San Juan River in Nicaragua, only to be invalided to Jamaica and then to England after almost dying of a tropical disease. Following convoy escort duties in the Baltic and the North Sea (1781) and along the Canadian and American coastlines (1782), he took advantage of the peace created by the conclusion of the American War of Independence to enjoy a three-month holiday in France (late 1783).

In 1784 Nelson sailed for the Leeward Islands in the West Indies, where he served for a three-year period marked by his strict enforcement of the Navigation Acts and the islanders' consequent hostility. After befriending Prince William Henry, the third son of King George III, and marrying the widowed Mrs. Frances "Fanny" Nisbet (on Nevis Island, 11 March 1787), he returned to England, where the Admiralty removed him from active service (along with countless others) to semiretirement and placed him on half pay.

After more than five years of semiretirement, he returned to active naval service in January 1793 when war with France became imminent. Serving in Adm. Samuel Lord Viscount Hood's Mediterranean fleet, Nelson witnessed the brutal capture of the French city port of Toulon by French revolutionary forces (December 1793), before fighting ashore as an artillerist on Corsica (January to August 1794). He contributed to the sieges of the Corsican seaports of Bastia and Calvi, losing the sight of his right eye during a bombardment near Calvi on 12 July. He nonetheless returned to sea soon after. On 13 and 14 March 1795 his ship *Agamemnon* engaged in a bitter fight off Genoa with the stronger French battleship *Ça Ira,* which established his reputation as a fierce and talented naval warrior.

He vastly enhanced this reputation at the Battle of Cape St. Vincent on 14 February 1797, when he helped Sir John Jervis (soon raised to the earl of St. Vincent) to defeat a superior Spanish fleet. Nelson was promptly made a knight of the Order of the Bath and promoted to rear admiral of the blue. His command of a detached squadron in July 1797 proved less successful; a failed amphibious attack on Santa Cruz de Tenerife in the Canary Islands cost him many casualties and his right arm as well, amputated above the elbow.

After recovering from this wound and his shattered confidence, in the care of his wife, Fanny, he returned to the Mediterranean in April 1798 with orders from St. Vincent to lead a squadron in search of a vast French invasion fleet headed from Toulon to an unknown destination. On 1 August 1798, Nelson's demoralizing three-month search ended; he found the French fleet at anchor in Aboukir Bay, near the mouth of the Nile in Egypt, and swiftly annihilated the French with no British ship losses. This victory, famous as the Battle of the Nile, made Nelson Britain's most popular hero.

He spent the following two years in Naples and Palermo, commanding various widely scattered British and Portuguese squadrons while at the same time meddling in central Mediterranean political affairs. The low point for Nelson came in June 1799 when he involved himself in both the recapture of Naples from Jacobin rebels and the execution and persecution of many of those rebels. He nonetheless returned to a hero's welcome in England in November 1800, after having traveled for three months through the German states with his new mistress, Emma Lady Hamilton, wife of British diplomat Sir William Hamilton. Nelson had encountered Lady Hamilton in Naples where he had first met her several years earlier.

Promoted to vice admiral of the blue in January 1801, Nelson soon found himself back in battle. His defeat of the Danes in the Battle of Copenhagen on 2 April 1801 was his second great victory. The British gained this victory only after Nelson ignored his commander in chief's ill-judged signal to withdraw at the height of the battle. The victory and Nelson's loyal disobedience increased his popularity with the British public to a fantastic level, and duly earned him elevation to Viscount Nelson of the Nile and Burnham Thorpe. The raids he ordered on French forces in Boulogne on 1 and 15 July 1801 proved unsuccessful, however, leaving him regretful that he had not led them himself. The Peace of Amiens (March 1802 to May 1803, preceded by a five-month armistice), which created an eighteen-month break in the war between France and Britain, gave him a chance to catch his breath and enjoy a relatively normal home life—with Emma Hamilton now residing with him in place of his wife—for the first time in many years.

All good things must end, they say. Nelson's did. Recalled to service in May 1803 and appointed commander in chief of the Mediterranean fleet in the flagship *Victory*, he commenced what became a two-year blockade of the French fleet in Toulon. Despite his masterful blockade, he failed to prevent Vice Adm. Pierre Jean Charles Baptiste Silvestre de Villeneuve from eventually breaking out in April 1805 as part of the French emperor Napoleon's plan to invade England. The French fleet crossed the Atlantic—with Nelson's own force in hot pursuit—before returning to unite with the Spanish fleet and take shelter at Cadiz. When the French and Spanish combined fleet put to sea, Nelson (who had received the briefest period of shore leave) caught it off Cape Trafalgar on 21 October 1805. His fantastic victory in this great naval clash claimed his life but earned him a true hero's immortality.

1

Nelson's Conception of His Enemies

NELSON GREW to crave Napoleon Bonaparte's demise and devoted ten years to that cause. He despised Napoleon, repeatedly describing him as a tyrant, an ogre, and a warmonger. He even wrote on one occasion that "like Satan," Napoleon was "a man of blood" and a "despoiler of the weak."[1] He also often confessed to "hating" the French people, whom he variously described as "pests," "vermin," and "a set of damned perfidious rascals."[2] He claimed that he learned such hate from his mother, who died when he was nine years old. For instance, he told Sir Gilbert Elliot on 8 October 1803: "I would not, upon any consideration, have a Frenchman in the Fleet, except as a prisoner." His explanation to Elliot was harsh: "You think yours [are] good; the Queen [of Naples] thinks hers the same; but I believe they are all alike. . . . not a Frenchman comes here. Forgive me; but my mother hated the French."[3]

Nelson was certainly not the only English person of his era to dislike the French. For centuries the two nations had suffered uneasy relations and occasional warfare. During Nelson's own infancy, they were locked in warfare in various parts of the world. By the last years of the 1700s, many ordinary English men and women wished for France's defeat. Their desire grew even stronger and more widespread after the collapse of the Peace of Amiens in 1803 placed Britain and France in a direct challenge between the two nations (separated only by the channel) for the first time.[4] Naturally, many of Nelson's naval superiors, peers, and subordinates also developed animosity toward

France throughout the 1790s and during the Napoleonic years. Nelson's friend Vice Adm. Cuthbert Collingwood, who served as second in command at Trafalgar, summed up this feeling in a letter to a relative. Why, Collingwood asked, has the coalition of so many countries against France not managed to convince the French people "that the plans of domination which their tyrants are pursuing are abhorred by all mankind?"[5] They must actually want, Collingwood concluded, to be ruled by these tyrants. One finds similar statements in the papers and diaries of many other soldiers, seamen, and civilians.

Yet, Nelson's countless expressions of animosity toward the French are remarkable for several reasons. First, the hatred they seem to convey is uncommonly intense and consuming, so much so that it almost overshadows the many positive aspects of the admiral's personality. Second, that hatred purportedly pertained to all French, not just the political or military leaders. Third, it pertained only to the French, and not to the other opponents Nelson faced throughout his career, including the United States, Spain, and Denmark.

Strong feelings against national opponents are natural during wartime, of course, although very few people ever get the chance to express their hostility through decisive action against their foes. Nelson did express his hostility by taking decisive action on many occasions, hence his suitability for analysis in this chapter. One might assume that his anti-Gallic loathing motivated him at each stage of his marvelous career from twelve-year-old midshipman to forty-seven-year-old hero and martyr, especially given his claims that he had hated the French since boyhood. He certainly considered the defeat of revolutionary and Napoleonic France and France's allies an exceedingly worthy cause, one for which he sacrificed much blood, an eye, an arm, and finally his life.

Through a study of Nelson's words and deeds, this chapter seeks to determine the role and significance of his claimed enemies in his worldview and thus shed a little more light on his motivations and goals. It argues that hatred of France played no role during his youth and early naval service. This hatred only emerged midway through his long career. And it was not based on any inherited dislike of the French people, their traditional social and political organizations, or their culture. It grew from several eventual realizations: first, that the revolutionary spirit that energized France as it swept away royalty and aristocracy was frightfully murderous and threatened what Nelson considered the basic foundations of society (God, divinely appointed monarchy, order, and social responsibility); second, that French military successes against Britain and its allies were humiliating and deserved revenge; third, that Napoleon had become the most powerful, celebrated, and feared warrior in Europe, which stung Nelson's own vanity; and finally, that the revolutionary and later Napoleonic French governments' bellicosity and desires for conquests seriously challenged Britain's commercial and colonial systems.

Early Hatreds

Nelson's naval career is easily divisible into three periods of frenetic activity, interspersed by quieter times, shore leave, and illness. These periods range from his appointment as midshipman in 1771 to his semiretirement on half pay in 1787, from his return to active service in 1793 to his defeat of the French fleet in Egypt in 1798, and from then until his defeat of the combined Franco-Spanish fleet off Cape Trafalgar in 1805.

The thirty-five years of Nelson's career were dramatic ones for Europe. The American and French Revolutions and the rise and conquests of the powerful dictator Napoleon ushered in new ideologies and constantly changed international borders, alliances, trade patterns, and national interests. When cast against the backdrop of these events, the three great periods of activity in Nelson's career make more sense and allow one to see how his conception of *the enemy* developed.

The dominant world event of Nelson's early career was the American War of Independence. This conflict pitted Britain militarily against its thirteen rebellious North American colonies from 1775, when opposing forces clashed at Lexington and Concord, to October 1781, when George Washington humbled the British at Yorktown. The war officially ended in 1783 with grudging British recognition of the new American nation's right to sovereignty and self-determination. One might assume that these dramatic events would have made an impression on Nelson, the patriotic young man whose first service in American waters preceded the war and who also served in the West Indies during some of its critical years.[6] The war certainly made an impression on Collingwood. He complained as early as March 1776 that King George III should make no accommodation with the American rebels and that the politicians in England who argued otherwise were "the supporters of a dangerous rebellion, rather than the asserters of publick liberty."[7]

Yet Nelson's surviving early letters, written when he was at sea and away from regular and current newspapers, reveal him to have been surprisingly *a*political, lacking interest in almost all political events that had no bearing on his own particular activities. To his "sea-daddy" William Locker, he did briefly express disappointment in October 1781 at the "sad news" that the Royal Navy had not prevented the French from landing reinforcements to help the American forces attacking Maj. Gen. Charles Lord Earl Cornwallis's British troops at Yorktown.[8] "I much fear for Lord Cornwallis," Nelson wrote, adding presciently: "if something [is] not immediately done, America is quite lost."

At that stage, however, the task-focused young captain was fully occupied preparing his frigate *Albemarle* for escort duties in the Baltic or North Sea and had little time to dwell on political events occurring thousands of miles away. Historians know of no Nelson letters commenting on the nature of American republicanism (probably because it resulted in few of the so-called "evils" of the later French variant, particularly atheism, popular tyranny, and territorial aggrandizement); on Cornwallis's subsequent surrender to Washington; or on the significant contribution France made to American victory.[9] Certainly, there are no exclamations of outrage or distress at any of these things.

Nelson did, though, eventually develop considerable hostility toward the new United States, and for a brief period in the mid-1780s, his hostility toward the nation he identified as an enemy reached a level almost approaching that which emerged in later years against the French. Yet his harsh criticisms of the Americans were not directed against the colonists' desire to rid themselves of British rule, their armed rebellion, or even their eventual success. The criticisms were directed against something far less worthy of Nelson's animus: the subsequent desire of the Americans to trade again with British colonies in the West Indies.

Particularly in his younger days, Nelson had a tendency toward dogmatism, and he tenaciously held out against other opinions on any issue once he had formed his own. This trait is nowhere more evident than in his service in the Leeward Islands squadron, based at St. Johns, Antigua, between 1784 and 1787. Technically, as a consequence of the war, former American colonists were no longer British subjects. They were considered foreigners and therefore, according to the Navigation Acts, not allowed to trade with the British colonies in the West Indies. Enforcement of the acts was not strict, however, and the colonies traded amicably with American merchants as needs demanded—*until Nelson arrived.*[10] Believing that British commercial interests must not be compromised by American traffic, and supported and encouraged by Collingwood and his younger brother, Wilfred, Nelson promptly commenced a rigid enforcement of the acts and even seized a number of American vessels.[11] This excessive zeal—perhaps best explained as a product of youth coupled with an excitable nature and initially unfamiliar responsibilities—earned him the ire of the island governors and most local islanders.[12] It also caused the Admiralty considerable awkwardness as it tried to arrive at a sensible resolution while defending Nelson's authority.

The situation sorely tested Nelson's patience, and he sent to friends in Britain a number of self-righteous and angry letters. The commander of his squadron and those around Nelson were "great ninnies" who clearly should know better than to turn a blind eye to—or "wink at" as he called it—the illegal American traffic.[13] Yet he directed his greatest hostility at the American traders themselves, treating them virtually as enemies and displaying an anti-American passion far exceeding anything revealed during the War of Independence itself. Nelson's vitriol even extended to islanders with supposed American leanings. They were equally disloyal, he exaggerated to his former captain William Locker, and were "inimical to Great Britain. They are as great rebels as were ever in America, had they [only] the power to show it. . . . I hate them all."[14] In another letter to Locker, he lamented "all the disasters" he had suffered "by having acted with a proper spirit against the villainies of a certain set of men settled in these Islands from America and [who] have brought their rebel principles with them."[15] These rebellious people, he wrote to the young widow Fanny Nisbet, were "trash."[16]

Interestingly, Nelson's accusations of "rebellion" relate to trade, rather than national, political, or ideological issues. Yet he used the most intemperate words to describe the situation. He told Mrs. Nisbet, his future wife: "I shall wish the [American] vessels at the devil and the whole Continent of America to boot."[17] When he noted on another occasion that he was looking after two seized vessels—or "Yankees," as he disparagingly called them—he added a wish that "they were a 100 fathoms under water."[18] He even insisted to Lord

Sydney, the secretary of state, that until "exemplary punishments" were meted out to those encouraging or allowing trade with Americans, "the evil never will cease."[19] If Lord Sydney considered Nelson's stance immoderate, he was not alone. Even Mrs. Nisbet, whom the angry captain would soon be writing to as "dearest Fanny," considered some of his actions unwise, especially the seizure of American ships. She told him so. "Duty is the great business of a sea officer," he indignantly replied.[20]

The young captain's excessive legalism, and his consequent strident anti-Americanism, eventually dissipated. Years later, as an admiral in the Mediterranean, he was even willing to offer Royal Navy protection to American merchant ships. But by then, of course, he had come to hate the French with almost volcanic passion.

The stridency of his anti-Americanism back in the mid-1780s should not be dismissed as unimportant. For several years it remained bitter, outstripping any feelings he had previously expressed about the War of Independence. More importantly, it also represented far greater passion than anything yet directed against the nation that he would later repeatedly and vehemently claim to have always hated: France. Indeed, during the entire first great period of his active naval service, right up to his reemployment in 1793, he remained surprisingly neutral in his political and moral views on France.

Even when Nelson had been in American waters or in the West Indies during the War of Independence, where naval competition between Britain and France was always heated, his diary entries, letters, and dispatches reveal no unusual hostility toward France for the assistance it provided to Britain's rebellious colonies. Moreover, his descriptions of Vice Adm. Jean Baptiste Charles Henri Hector Comte d'Estaing's powerful French fleet in the West Indies, which in 1779 posed a grave danger to the British colonies, are bland and without the "may they go to the Devil!" and "may we teach them a lesson" statements that would later be peppered throughout his correspondence.[21] He merely stated that should the comte's fleet gain victory, he would doubtless soon be speaking French (presumably in captivity).[22]

Other occasional colorless references to the French as "the enemy" and descriptions of engagements or ship sightings are the only indication from Nelson's writings before the middle of 1783 that he gave any thought to the French and their international activities. Then, in October of that year (1783), only six weeks after the Treaty of Paris officially brought peace between the

United States, France, Spain, and Britain, Nelson even traveled to St.-Omer, France, for a three-month holiday. He did so ostensibly to learn French, a commercially and socially useful language that was also used in royal courts throughout Europe.[23] The French were apparently no longer his enemy.

As he traveled through France, Nelson was struck by the poverty of the peasantry, which seemed far more widespread than it was in England. Yet this realization, and his complaints about one inn he and his companion stayed at—he called it "a pigstye" and lamented, "O what a transition from happy England"—belie the fact that he actually enjoyed France and marveled at the stunning beauty of its countryside. He described a trip to Montreuil in these terms: "[We] passed through the finest corn country that my eyes ever beheld, diversified with fine woods, sometimes for nearly two miles together, through noble forests. The roads mostly were planted with trees, which made as fine an avenue as to any gentleman's country seat."[24] He found St.-Omer equally lovely and commented on its cleanliness and its attractive, well-paved and well-lit streets. This circumstance surprised him as he had been given to believe it was a "dirty, nasty town."

Nelson also found the people of St.-Omer relaxed and very pleasant, especially the young women. The family he lodged with had two pretty daughters, which gave the sea captain further incentive to master their language. "I must learn French if 'tis only for the pleasure of talking to them, for they do not speak a word of English." His infatuation with the French girls dissolved after he met an English clergyman who had, as Nelson told his brother William, "two very beautiful young ladies, daughters. I must take care of my heart, I assure you."[25] Yet, even after returning to England, he fondly remembered all the "charming" French women he had met and "would like to return to."

Nelson's letters reveal only happy experiences in France, which leaves a comment to William, written a few weeks after he returned to England, hard to understand. "I want to be proficient in the [French] language," he wrote on 31 January 1784, a full five years before the revolution commenced, but "I hate their country and their manners."[26] Clearly he had forgotten that only a month before, he had written from St.-Omer to the same brother: "At this place I am as happy in a friendly society as is possible."[27]

The out-of-place expression of hate for the French may stem from his realization, once back in England, that his future employment in meaningful roles—in other words, in the exhilarating naval roles he loved—depended on renewed hostilities with a major nation, and that most likely meant France. This view certainly explains his comment to William Locker, written four months after his return from France: "I thank you much for your news, which if true, [means that] hostilities must commence soon again with the French; God send, I say."[28] At that stage Nelson was in command of the frigate *Boreas* and was preparing to carry Lady Jane Hughes and her family to the West Indies, where her husband, Rear Adm. Sir Richard Hughes, was commander in chief of the Leeward Islands station. This was not an exciting task by any standards, much less for someone who craved action and glory and had wanted to go to the *East* Indies.[29] And, as outlined previously, it all turned out poorly; Nelson began upsetting people in the Leeward Islands almost as soon as he got there. His overzealous enforcement of the Navigation Acts not only alienated the islanders, leaving him lonely and bitter, but also doubtless contributed to the Admiralty's decision to include him in its list of captains to be placed on half pay. Nelson ended up spending over five years, from July 1787 (the *Boreas* was finally paid off in November) to January 1793, back home in Norfolk without another sea-going appointment.

Toulon and After

"Post Nubila Phoebus"(after clouds comes sunshine)! Nelson excitedly penned this phrase to Fanny, his wife of almost six years, on 7 January 1793, the day after he learned that his time in the wilderness had ended and that he was about to receive command of another ship.[30] The second great period of his career was about to begin, a period dominated by the tumultuous French Revolution and its often cruel and savage consequences. Nelson's first direct exposure to all this, and the destructive impact it had on British diplomatic and commercial relations with allied nations, not to mention on the reputation of his beloved Royal Navy, planted in him a hatred of the French that quickly grew far more intense than any emotion hitherto directed at a people or nation.

Nelson had not done anything new or different to deserve his return to favor with the Admiralty. His reemployment, and the subsequent hardening of his attitudes toward France, resulted directly from events happening on the other side of the channel. Since the signing of the Treaty of Paris on 3 September 1783, Britain and France had enjoyed a period of peace and relatively amicable trade relations. Nelson was not alone in being unemployed; scores of fellow officers and countless seamen also found themselves lacking work. This peaceful environment was coming to a close by decade's end, however, and by 1789 war between the two nations seemed inevitable. In May of that year, the French king, Louis XVI, responded to what appeared to be looming bankruptcy by calling a meeting of the Estates-Général (the first since 1614). On 17 June 1789 the Third Estate, joined by reformers from the nobility and clergy, declared itself to be a National Assembly and three days later swore never to dissolve until France had an American-style constitution. The Crown promptly nullified these declarations and ordered the National Assembly to rearrange itself into the original three estates. It refused.

Other events quickly overtook this standoff. On 14 July 1789, after two days of riots and destruction, an angry Paris mob stormed and occupied the prison-fortress of the Bastille, considering it symbolic of oppressive rule. Order collapsed and revolution soon spread across the country. Privilege was abolished, the king became a constitutional sovereign, and a more equitable distribution of wealth was promised. Yet this was not the end. Neither Louis XVI nor his nobles were prepared to accept the new order—or *dis*order as it really was—while disagreements between various revolutionary factions over the nature and extent of the changes they wanted continued to boil over. In the ensuing struggles throughout the next few years, countless nobles fled the country and soon formed a hostile, menacing force of émigrés in England and the nations bordering France. Then, on 21 January 1793, four months after abolishing the monarchy and declaring France a republic, Jacobin revolutionaries shocked most other Europeans, who followed events through newspaper reports, by having King Louis XVI executed by guillotine.

Britain's preparations for war had intensified rapidly after France's victory over the Austrians at Jemappes on 6 November 1792, which swiftly led to the lightning occupation of Belgium by the French. Britain had been dreading this occupation, seeing it as France's first step to gaining dominance over Britain's important new ally, Holland. Prime Minister William Pitt had declared in 1789, for instance, that the need to keep Belgium independent of France was "worth the risk or even the certainty" of war.[31] Now that the dreaded event had become a reality, Britain began sending blunt warnings to France that intervention in other nations' affairs, particularly Holland's, would result in war. Finally convinced that France would not back down, the British from December 1792 on began maneuvering the French into declaring war first, a goal successfully attained on 1 February 1793, ten days after Louis's execution.[32]

Nelson had been closely following these events in France, as had most of his peers, including Collingwood. The latter, for instance, suspected the French had gone mad and doubted that "such an ungoverned mob" would be able to reestablish orderly government.[33] Yet, unlike Collingwood and others who had already returned to active naval service, Nelson's primary interest was not the suffering of ordinary French people, although he sympathized with their plight. Neither was he interested in the political intrigues and leadership struggles in Paris, which of course upset his monarchical, conservative sensibilities; nor even in the revolution's avowed atheism, which offended his deeply rooted Christian beliefs. He was interested mainly to see whether the French situation would blow up and result in war, which might necessitate, he dearly hoped, his reemployment. "Is there any idea of our being drawn into a quarrel by these commotions on the Continent?" he asked William Locker in September 1789. "Whenever that may be likely to happen I will take care to make my application in time."[34]

Like other captains who found themselves needed again in 1793, Nelson was delighted to learn that war had given him another chance in the Royal Navy. He threw himself with the utmost energy into his new duties, preparing his vessel, the sixty-four-gun *Agamemnon* (which was also brought out of retirement) to join Admiral Lord Hood's strong fleet in the Mediterranean. There it would operate off the shores of the Maritime Alps and Genoese Riviera in support of the Austrian army's campaigns against the French.

Hood surprised the British government in August 1793 by accepting custody of Toulon, France's great southern port, at the invitation of local French royalists.[35] They had already suffered dreadfully under the Terror and now hoped the British and Spanish forces under Hood would protect them from their revolutionary countrymen. Toulon's defense would be difficult, Hood knew, but he immediately began raising a force of Britons, Spaniards, Neapolitans, Sardinians, and French royalists, not all of them combat worthy. Nelson, who sailed into but never actually set foot in Toulon himself, was involved in the raising of these troops; and for his ceaseless efforts, the energetic Nelson earned Hood's praise. Yet Toulon could not be held. The French leadership in Paris sent an unknown twenty-four-year-old captain (hastily promoted to lieutenant colonel) named Napoleon Buonaparte—he dropped the *u* in his name after 1796—to take control of artillery besieging the city. Although he initially clashed with Jean-François Carteaux, the general in charge of the overall French assault, Napoleon's masterful use of artillery soon rendered Hood's defense of Toulon untenable.[36] By clever use of his cannons, Napoleon demolished a key fort and began bombarding the British fleet, which hastily embarked its troops and as many of Toulon's terrified inhabitants as it could and then put to sea. Four days later, on 22 December 1793, the victorious young Bonaparte was promoted to brigadier general.[37]

Nelson was aghast at the port city's fate. He had been euphoric only four months earlier when Toulon, "the strongest place in Europe," had placed itself in British care without a single shot being fired. It was such an event, he told his wife, that "history cannot produce its equal."[38] Now he was stunned to learn this unprecedented event had ended with a humiliating defeat, the evacuation of the Royal Navy, and the unjustifiable massacre of over a thousand remaining royalists (to which Napoleon turned a blind eye and may even have contributed).[39] Nelson explained away Toulon's recapture as a case of overwhelming French superiority, repeating wildly exaggerated reports of sixty thousand French troops storming the city.[40] But he was especially bothered by the damage it would do to British morale. He also realized it would give an immense boost to revolutionary French spirits.[41]

He did not mention Napoleon personally and may not then have known who caused Toulon's recapture. Yet the event itself and the consequent atrocities did much to convince Nelson that the Jacobin revolutionary spirit that had swept through France, proclaiming liberty and equality but causing widespread misery and tyranny, was quite unlike the American revolutionary spirit of freedom and self-determination. The Jacobin scourge could not be allowed to flourish. It had to be destroyed. Indeed, the events in Toulon can be seen as a turning point in Nelson's attitude toward the French.

Back in August 1793, Nelson had been expressing disapproval of the widespread fear and chaos in France—where only two types of people apparently now lived—"the one drunk and mad; the other, with horror painted in their faces, and absolutely starving"[42]—and noting with displeasure that "the guillotine is every day employed." Yet, perhaps because reports of these events lacked immediacy and because the events themselves were not occurring in his own area of responsibility, he saw the turmoil as largely a matter of French domestic politics and questioned the real need for such a massive British response. He even wrote to his wife: "I hardly think the war can last, for what are we at war about?"[43]

However, Toulon's recapture and the vengeful slaughter of its remaining royalists deeply shocked Nelson and virtually everyone in Hood's fleet.[44] Nelson told the duke of Clarence that the "scene of horror" was enough to break even "the hardest heart."[45] Similarly, he mourned to his wife: "All is horror, . . . I cannot write all, my mind is truly impressed with grief. Each teller makes the scene more horrible."[46] Nelson felt genuine and profound horror unlike his friend Collingwood, who was in England during this period and consequently referred to the recapture of Toulon only in terms of strategic mismanagement, with no evident grief at the atrocities.[47] Nelson's letters and dispatches subsequently became far more bitter toward revolutionary France, with expressions of disgust at its violence along with yearnings for France's destruction appearing with increasing regularity.

By mid-1794 he was even criticizing Britain's principal allies, especially the Hapsburg Empire (normally known simply as Austria), which was the strongest Continental member of the First Coalition. Austria and the other allies, Nelson said, were forever squabbling among themselves and not making an adequate effort toward the "common cause" of defeating all French forces outside the new republic's borders. "The Allied Powers seem jealous of each other," he wrote to his wife on 27 September 1794, "and none but England is hearty in the cause."[48] Two weeks later he complained to William Locker: "Our Allies in these seas are useless to us, as I believe they are in all other parts of the world."[49] He became so frustrated by what he considered the inactivity of Britain's allies that during occasional moments of pique, he even stated that Britain should withdraw from the coalition and make a separate peace with the French, who could continue killing themselves for all he cared, and leave the Mediterranean allies to fend for themselves.[50] "I wish we could make a Peace on any fair terms," he wrote to Fanny on 24 October, "for poor England will be drained of her riches to maintain her Allies, who will not fight for themselves." He told another correspondent: "Our Allies are our burden. Had we left the Continent to themselves, we should have done well, and at half the expense."

The year 1794 was indeed hard for the British. Their success in Corsica —the battles in which Nelson was heavily involved (costing him the sight of one eye)—and in the Atlantic, where Vice Adm. Richard Earl Howe gained a pleasing naval victory (which became known as the Glorious First of June), failed to compensate for misfortune elsewhere. The French drove British, Austrian, and Dutch armies out of France and Belgium; pushed the Austrians and Prussians back to the Rhine; and advanced, driving Spaniards before them, through the Pyrénées. The only allies left unbroken were the Italian states, which looked vulnerable to a French onslaught. Through all these setbacks, which he witnessed or read about in the newspapers he frequently received, Nelson cursed the French, who "behave famously ill, not like men of spirit,"[51] and worked tirelessly to thwart their ambitions in his own area of service. He proved useful during combat on land before returning to sea in late-December 1794 as part of Adm. William Hotham's force blockading Toulon (Hotham had replaced Hood in September).

The return to sea should have eased at least some of Nelson's frustrations. He no longer had to witness French revolutionary atrocities and what he considered the half-hearted efforts of the British and other armies, and he was finally getting the chance again to hurt the French where he believed them most vulnerable: at sea. In fact, he sometimes wished that all British troops were withdrawn from the Continent, "and only a naval war carried on, the war where England alone can make a figure."[52] Yet even at sea, where newspapers reached him less often and long after the events described, Nelson lamented at reports of republicanism's apparent spread across western and southern Europe. The supplanting of responsible administration and government by atheistic republican "mobs," he believed, resulted in hunger, poverty, despair, and random and uncontrolled outbursts of anger and violence.

He consequently remained remarkably bitter at the French and became increasingly annoyed by Hotham's lack of aggression toward them. After two days of inconclusive battle with a French squadron in March 1795, for instance, he complained about his admiral's unwillingness to pursue the French squadron. Fourteen British ships had engaged seventeen French, and captured only two, yet Hotham said, "We must be contented. We have done very well."[53] Nelson hated this lack of fighting spirit and told his wife that "had we taken 10 sail and allowed the 11th to have escaped, if possible to have been got at, I could never call it well done."[54] Three months later Hotham refused to let him attempt to take two French frigates. Thankfully, Nelson noted to Fanny, two other British captains found and took them in a spontaneous attack during a separate mission. He was relieved and wrote, "Thank God the superiority of the British Navy remains, and I hope ever will. . . . Had our present Fleet but one good chance at the Enemy, on my conscience, without exaggeration, I believe that if the Admiral would let us pursue, we should take them all."[55]

Throughout the rest of 1795, Nelson worried gravely about the security of Britain's weaker Mediterranean allies.[56] He therefore continued to rail against French anarchy, wanton violence, and military conquests as well as what he considered the passivity of the powerful Austrians. He was particularly angry at the Austrian army, which he and a small squadron supported as it attempted to drive French forces from the Italian Riviera and so gain control of vital north Italian ports. Complaining in October 1795 that the campaign was petering out, Nelson added sarcastically that he couldn't tell it had ever started. In any event, he added, the allied efforts were so miserable that a conclusion to hostilities depended entirely on the will of the French themselves, who might yet decide to bring back monarchical government and end the war.[57] Nelson had even become so embittered by the failure of French royalists to unite and fight the republicans, and by the financial costs the former imposed on the British, that he sometimes sourly said he did not care whether the monarchy was reestablished. A monarchical Frenchman was no better than a republican. "To me," he insisted to William Suckling (a close relative by marriage), "all Frenchmen are alike; I despise them all. They are (even those who are fed by us) false and treacherous."[58]

Nelson's disgust at the Austrians—"thousands" of whom "ran away who had never seen the enemy"[59]—and his hatred of the French reached new heights after their army achieved a powerful victory over the Austrians in the Battle of Loano on 23 November 1795. Even though he grudgingly admitted to admiring the French tenacity and fighting spirit, noting they had gained victory despite being starving and poorly clothed, he placed most blame onto Britain and its "miserable" allies. Their ineffectiveness and lack of fighting spirit had given revolutionary France a victory and a huge morale boost and thereby had ended any likelihood of its return to monarchy. "*We* have established the French Republic," he glumly told a friend.[60] The "volatile, changeable" French might well have reverted to orderly government, but it was too late now. This gnawing sense of guilt made him more bitter against the French —*all* French. "Whether Royalists or Republicans," he said, they remained equally worthy of his "detestation."

Detesting the French throughout 1796 proved easy, especially after Napoleon, whose name was quickly rising to fame in France and infamy elsewhere, assumed command of the French-manned Army of Italy in March and began thrashing the Austrians and their allies.[61] It seemed to Nelson that British action in the Mediterranean was a constant series of disappointments, with the main exception being Hotham's replacement by Sir John Jervis. Hotham's perceived lack of aggression against the French had continued to frustrate the anti-Gallic and glory-seeking Nelson, who insisted in one letter home:

> If we are to finish the war with France, we must not be disposed to stop at trifles; it has already continued much too long. . . . We have much power here at present to do great things, if we know how to apply it. Hotham must get a new head; no man's heart is better, but that will not do without the other.[62]

Jervis, on the other hand, was the type of commanding admiral that Nelson had always wanted: intelligent, aggressive, resolute, anti-French, and openly appreciative of Nelson's emerging genius for war.

After Napoleon's army occupied the port of Leghorn in late June 1796, thus denying the British fleet one of its last ports in the Mediterranean, Jervis sent Nelson and his squadron to blockade it. The commodore, as Jervis had named him, relished the chance to hurt the French and made the most of it. Nelson despised Napoleon (who was "no gentleman") for the brutal and ultimately murderous way he reportedly treated Leghorn's political leaders, and for replacing them with his own "tyrannical" team.[63] Nelson thus implemented an extremely tight blockade of Leghorn, even though he felt sad at its impact on ordinary Leghornese, for whom he had nothing but "good will."[64]

Nelson also captured Porto Ferraio, on the Island of Elba, and Capraia, off the north coast of Sardinia, to eradicate piracy and reduce Napoleon's chances of retaking Corsica. He was delighted to be active and successful against the enemy he hated. And hate the French he now did. After confessing to Collingwood that he sometimes wished the Leghornese would "cut the throats" of their French overlords, he added: "What an idea for a Christian! I hope there is a great latitude for us in the next world." He closed his letter with a "most sincere wish for driving the French to the Devil."[65] Nelson consequently feared that his separation from Jervis's main fleet might rob him of the opportunity to destroy the French at sea. "I have only now to beg," he wrote to Jervis, "that whenever you think the Enemy will face you on the water, that you will send for me; for my heart would break to be absent at such a glorious time."[66]

The "glorious time" would come, but not in 1796. In fact, that year closed "shamefully" for Nelson, Jervis, and the British fleet. In October, Napoleon's marvelous successes in Italy prompted Spain, which was already suffering defeats and risking invasion, to change sides and proclaim war on Britain.[67] The British government responded by declaring its position in the Mediterranean logistically untenable and instructed Jervis's fleet, after evacuating British troops from Corsica, to leave the Mediterranean altogether. Nelson was more disgusted than he had ever been at a British strategic decision. He felt "grieved and distressed [and] ill" at Britain's humiliation and that its allies would be defenseless,[68] Napoleon's coastal lines of communications would be unhampered, the Mediterranean would become a virtual French lake, and the Royal Navy would probably be unable to hurt the French anywhere for quite some time. He told his wife,

> We are all preparing for an evacuation of the Mediterranean, a measure which I cannot approve. At home they know not what this Fleet is capable of performing; *any and everything*. . . . I lament in sackcloth and ashes our present orders, so dishonourable to the dignity of England, whose fleets [*sic*] are equal to meet the world in arms.[69]

Jervis wasted little time in avenging this indignity, however. On 14 February 1797, he caught the numerically superior Spanish fleet off Cape St. Vincent and, thanks to the initiative displayed by Nelson in particular (but by others as well) and the superiority of British captains and crews, secured a much-needed victory. Four Spanish ships were captured, two by Nelson himself (although clearly with assistance from other ships). Britain rewarded Jervis by raising him to the peerage as admiral the earl of St. Vincent and Nelson by promoting him to rear admiral (which was due anyway) and making him a knight of the Bath.

Delighted though he was, these honors could not diminish Nelson's desire to gain revenge over the nation that had recently forced his beloved Royal Navy from the Mediterranean—which was *not* Spain, but *France*. Yet it would be some time before he would get an opportunity to punish the French. In the meantime, St. Vincent sent him and a detached squadron on a mission against the Spanish, whom Nelson always considered more of a nuisance than a competitive rival power. His mission was not even against a Spanish fleet or squadron, but was an amphibious attack on the town of Santa Cruz de Tenerife in the Canary Islands, where a treasure ship was supposedly anchored.[70]

The attack in July 1797 proved a miserable failure—the worst in Nelson's distinguished career—and cost him 343 casualties killed or wounded and the loss of his right arm, amputated after receiving a shot through the upper arm.[71] Still, Nelson accepted his defeat as God's will and as a lesson in humility. He also displayed a sensitivity of spirit toward the Spanish victors, led by Don Antonio Gutiérrez, that matched the generous and respectful way he had treated defeated Spaniards at the Battle of Cape St. Vincent, but which he probably would never have displayed toward a Frenchman in similar circumstances.

Recovery from his amputation was slow, painful, and frustrating, as was his acceptance of the collapse of the First Coalition on 17 October 1797, when France and Austria concluded the Treaty of Campo Formio. France gained from Austria the Low Countries, Venetian Albania, and the Ionian Islands. The Cisalpine Republic (the pro-French northern Italian republic created by Napoleon) gained Milan, Mantua, and Modena. Britain was now as alone in its war against revolutionary France as it would be late in 1940 against Hitler's evil Reich.

The one-armed Nelson found this all bitterly distasteful, especially as the Austrians blamed their reverses on the withdrawal of British naval support in the Mediterranean. Considering this an attack on Britain's and the Royal Navy's honor, he fervently sought to return to active service against the French. His requests to the Admiralty paid off. He was back in action in March 1798, when he hoisted his flag on the seventy-four-gun *Vanguard*. The earl of St. Vincent was about to give him the mission of which he had often dreamed: an attack on a French force commanded, or at least headed, by the French general whose victories in Italy had given him an international reputation for military genius—Napoleon.

Napoleon's build-up of an invasion fleet in Toulon and Genoa had alarmed the British government, which feared that the Royal Navy's total absence from the Mediterranean left Napoleon with a free hand, possibly for an attack on England or Ireland. The Admiralty therefore instructed the earl to take his fleet or send a detached squadron back into the Mediterranean in order to reconnoiter and thwart the French plans, whatever they were.[72] In a remarkable gesture of respect and trust, the earl chose Nelson above more senior officers for this most critical mission.[73]

From leaving Gibraltar on 8 May 1798 until the destruction of Napoleon's Egyptian invasion fleet three months later, Nelson thought of almost nothing else but finding his enemy and bringing him to battle.[74] He hungered for glory, but was also acutely aware that his failure to thwart Napoleon's ambitions could end in national disaster, especially given the grave threat that Napoleon's fleet posed to critical British interests in Portugal, Ireland, the West Indies, Italy, and even India. The responsibility weighed heavily upon Nelson; this was his first proper fleet command (in reality it was still a detached squadron, albeit an independent and powerful one), the eyes of the Admiralty were upon him, and his nation craved a victory over the "brutal and ungodly" French.

Even before he reentered the Mediterranean after almost eighteen months' absence, Nelson felt depressed that the French not only dominated the Continent, but also now ruled that great sea where the Royal Navy had no presence and power.[75] He worried especially about his inability to learn Napoleon's plans. Since the Royal Navy had left the Mediterranean, its network of contacts had diminished, and this, coupled with an absence of frigates in that sea, left the Admiralty with a severe lack of naval intelligence. Nelson knew a powerful enemy fleet was sailing somewhere but could only guess at its likely destination.

Nelson initially worried that Napoleon, whose popularity in France rivaled Nelson's own in England (a fact that clearly annoyed him), would not personally accompany the French fleet.[76] Yet, after learning Napoleon would join the fleet, Nelson rejoiced at the unprecedented opportunity for glory now presented to him. Napoleon had become the most famous and feared man in Europe, but no British commander had yet challenged him head-to-head on land or at sea. Nelson desperately hoped he would be the first. And if he could only bring the Corsican to battle, Nelson would destroy Napoleon's grand reputation while vastly raising his own. It would be his "delight," he told his wife on 20 July 1798, to meet Napoleon "on a wind, for he commands the fleet as well as the army. . . . Glory is my object, and that alone."[77]

Despite this flamboyant claim, Nelson was not only pursuing personal glory, trying to reduce that of his counterpart Napoleon and seeking revenge for recent British setbacks and French atrocities—he harshly described the French as "these pests of the human race"[78]—he was also striving to prevent France, Britain's greatest economic and military rival, from extending its already substantial influence in the Mediterranean or striking a serious blow to British trade, commerce, and empire anywhere else. He was desperate to stop Napoleon and, after his first month of fruitless searching, confessed to being "in a fever" of anxiety.[79] He cursed the French for eluding him, telling Sir William Hamilton, British representative at Naples and husband of Emma (soon to be Nelson's mistress), that "the Devil's children have the Devil's luck."[80] Yet he knew he would eventually find and attack them—hopefully on the open sea, but ultimately wherever they were and regardless of relative strengths or situations.[81] "You may be assured," he told the earl of St. Vincent at one point, "that I will fight them the moment I can reach their fleet, be they at anchor or under sail." Even if the French fleet was secure in a strong port, he told another friend, he would still attempt its immediate destruction.[82]

Nelson lived up to his word. Early in the afternoon of 1 August 1798, he finally found the French fleet, which was not at sea with Napoleon in command as Nelson would have preferred. It sat anchored in compact order of battle in Aboukir Bay, Egypt, and Napoleon's Armée d'Orient had clearly departed for battle inland. Still, Nelson finally had a French fleet in his grasp, and, with Napoleon present or absent, he was determined to send it to hell.

It was almost 1730 hours, and darkness was fast approaching by the time Nelson's fleet closed on the French and formed its own line of battle. Few naval battles had ever occurred at night, and the French commander, Adm. François Paul Comte de Brueys d'Aigailliers, doubtless assumed that any reasonable foe would wait until morning to launch an attack rather than risk a night battle on an unfamiliar coast, which could result in "friendly fire" incidents and groundings on dangerous shoals. Yet Nelson, always unorthodox, was unwilling to forfeit the element of surprise or let the French fleet slip away under cover of darkness. He attacked immediately. The battle was bitter, the result inevitable. Dawn revealed a British victory of unimagined totality. Indeed, it was the most devastating naval victory of the eighteenth century. For no ships lost of their own, the British had virtually annihilated the French fleet, with only two of its thirteen battleships escaping capture or destruction. It was a crushing defeat for the absent Napoleon, whose army was now stranded in Egypt, and a hard blow for the French Republic, which lost its naval mastery in the Mediterranean.

Nelson was naturally ecstatic, telling Fanny: "Victory is certainly not a name strong enough" for this success.[83] He felt convinced that God's "almighty hand has gone with us to the battle, protected us, and still continues destroying the unbelievers."[84] He also concluded that he had humbled Napoleon in a way that no one yet had (as Nelson had dearly hoped) and that he had established the right preconditions for the "dangerous" general's total defeat. He had destroyed the French fleet, returned mastery of the Mediterranean to the Royal Navy, and trapped the Armée d'Orient in the heat and hostility of Egypt, where it would receive no reinforcements and logistical support. Nelson had little doubt that Napoleon's army would be "destroyed by plague, pestilence, and famine, and battle and murder," adding: "that it may be soon, God grant."[85] He also believed that Napoleon's severe misfortunes in Egypt would encourage Britain's allies to renew hostilities against the French, "for until they are reduced there can be no peace in this world."

The news of Nelson's victory at the Nile was greeted with true delight throughout Britain and its colonies, and with total euphoria at the Court of Naples, whose king, Ferdinand IV, decided he would indeed take the offensive against the French.[86] With encouragement from Nelson, he entered Rome on 29 November 1798 and drove out the French.[87] But his offensive was misguided, given that his army lacked experience and resolve and was not receiving the anticipated support of major allies. A serious reversal of fortune soon occurred: the French retaliated and marched south toward Naples, with Ferdinand's army quickly breaking. Things went better elsewhere, though. Nelson's victory at the Nile inspired the Ottomans to make war on Napoleon's trapped army in Egypt and also impressed the Russians, who agreed to enter into understandings (and later a coalition) with the British government.[88]

From the Nile to Copenhagen

Nelson would never again chase or face a fleet or army under Napoleon's direct command, much less engage and destroy one. That honor would go to the splendid duke of Wellington, a decade after Nelson's death off Cape

Trafalgar in October 1805. Yet the admiral's contribution to Britain's security and successes during the period from the Battle of the Nile to his untimely death seven years later—the third great period of his career—was truly matchless. It outstripped even Wellington's marvelous achievement at Waterloo. And throughout those seven years, despite being distracted by his torrid affair with Emma Lady Hamilton, which inspired some poor combat decisions, and by his diversion to the Baltic, where he won a fiercely contested battle against the Danes, he seldom placed other priorities ahead of Napoleon's defeat.

After his Egyptian victory Nelson chose not to return to his wife in England. Instead he traveled to the Kingdom of the Two Sicilies, ostensibly to have his flagship repaired in the shipyards at Posillipo, but probably also to recover from almost total exhaustion and a forehead wound in the company of people who would shower him with the praise he so enjoyed.

Thus began a two-year period in Naples, during which he expertly commanded a large force comprising British and Portuguese squadrons scattered as far afield as Acre, Alexandria, Malta, Naples, and the Gulf of Genoa. He nonetheless sometimes found it hard to turn his gaze from Emma Hamilton and the Court of Naples. Dazzled by royal favor and caught up in local and regional politics, Nelson even encouraged the premature military offensive of King Ferdinand, whom he served with more fervor and adoration than did almost any Neapolitan.[89] And when the French counterattacked and swarmed into Naples, Nelson's ships provided the only hope of salvation for Ferdinand and his court. Abandoning mainland Italy, they fled to Palermo, their second capital, on the northwest coast of Sicily.

In February 1799 Ferdinand's royalists finally began to fight back against the French occupation of their Italian possessions, with Cardinal Fabrizio Ruffo leading an ill-disciplined army that committed many outrages as it slowly reclaimed territory. Nelson, who considered Palermo "detestable" but wouldn't remove himself from the royal family, felt bad about the loss of Naples.[90] He was determined to do all in his power to see it back under Ferdinand's rule, even if he should die in the attempt.[91] He assisted by blockading Naples and conveying a Neapolitan royalist army to help Ruffo retake the city from local republicans who had sided with the French.[92] The royalists triumphed in June. King Ferdinand's vengeance proved terrible, with atrocities and executions commonplace after the city's recapture.

Nelson was himself merciless toward what he and his circle called "the

Jacobins," thereby associating them ideologically and morally with early French republicanism. They were, in fact, local revolutionaries who had temporarily transformed southern Italy into a French-sponsored republic. Unreservedly endorsing Ferdinand's brutality, Nelson used his squadron to prevent republicans withdrawing to Toulon or elsewhere in France, and even had them handed over for execution. He discouraged appeals for clemency, insisting that "rebels, Jacobins and fools" deserved no mercy.[93] His refusal of clemency even extended to Commo. Francesco Prince Caracciolo, who pleaded unsuccessfully for an honorable death by firing squad, as befitting his status and high rank, rather than being "hung up like a felon and a dog."[94] Nelson's refusal prompted an otherwise devoted naval lieutenant to later write that, despite "poor human nature" being fallible, Nelson "sinned, and deeply sinned."[95] He did this in violation of a truce organized earlier by one of his captains, Edward Foote, which he had promptly and harshly overturned.[96]

His support of the brutal and often murderous persecution of republicans has caused much debate among Nelsonian scholars, with even many of his hagiographers finding it hard to explain away. By today's standards—indeed, by almost any standards—his behavior would be considered unnecessarily cruel. Yet, without trying to minimize the severity of his actions, it is possible to see them as a logical consequence of his experiences with French republicanism during the previous decade. As noted, he had witnessed or read about the worst aspects of the French Revolution: regicide, the Terror, godlessness, widespread anarchy, hunger, and atrocities. He simply couldn't bear to see the same things spread elsewhere, especially if they destroyed the prosperity, happiness, and security of his dearest friends in Naples. Those friends included Queen Maria Carolina, who had herself developed a pathological hatred of Jacobins and all other revolutionaries. This is understandable: her sister, Queen Marie Antoinette of France, had been infamously murdered by guillotine in 1793.

Nelson feared that the French evils of 1793 were starting to visit the Kingdom of the Two Sicilies, and wanted them stopped before it was too late. He was not alone in feeling this way; Capt. Alexander Ball, one of his immediate subordinates, wrote in a letter to Lady Hamilton: "I cannot express my astonishment and sorrow when I first heard of the political revolution which the Jacobins have effected [in Naples]."[97] He added, using a metaphor that Nelson and other anti-Jacobins also used, equating republicanism with disease: "I hope the Sicilians will not catch the infection."[98]

Nelson considered the French solely to blame for the spread of this contagion. He told Sir John Acton, the Neapolitan prime minister, that republicanism "is the system of *terror*, by which terror the French hold all Italy."[99] To call republicans "patriots," he raged, was "a prostitution of the word!"[100] In short, Nelson possessed, as one document from that period states, "an implacable hatred against all revolutionists of the French type," and believed that "it was necessary to cut out the root of the evil so as to avoid new misfortunes."[101]

Nelson's actions were not only brutal, they were also of dubious legality. He had certainly received no instructions from the Admiralty or any other British authority to involve himself in these matters, no matter how justified by circumstances he personally considered his decisions. In July 1799 he exacerbated the situation by defying an order from Vice Adm. George Lord Keith (St. Vincent's successor) to concentrate his forces around Minorca, which Keith hoped to protect from an invasion by a French fleet that had recently entered the Mediterranean. Nelson agreed to send some ships but claimed, perhaps no longer being able to see the wood for the trees, that he had to remain with the other ships under his command to protect Naples and Sicily.[102] This failure to obey orders was of a different magnitude than his earlier spontaneous demonstrations of initiative during battle and earned Nelson a stern rebuke from the Admiralty.[103]

Throughout his time in Naples, he kept busy helping to govern the kingdom, a task for which he was ill suited, while letting some world affairs pass him by. But by March 1799 Nelson had become focused on cleansing Italy of French influence and on defeating the French garrison in La Valetta, Malta, despite insisting that he remained committed to "the destruction of that robber, Buonaparte, with all his gang," to the head of the Ottoman squadron in the Mediterranean.[104] Uncharacteristically, Nelson let others, particularly Sir Sydney Smith (a newly arrived captain who had been forced to accept Nelson's overall authority), worry about Egypt and the Levant.

Nelson followed reports of Napoleon and his army in Egypt, whose "extirpation" he claimed to long for.[105] And he naturally rejoiced at news that the famous French general had, due to Smith's stout defense, reluctantly lifted his three-month siege of Acre. He was returning to Egypt, leaving behind his cannons, siege weapons, and sick. "That vagabond has got again to Cairo," Nelson told his friend Thomas Troubridge, "where I am sure he will terminate his career. . . . Adieu, Mr. Buonaparte!"[106] Yet Nelson was so immersed in Neapolitan, central Mediterranean, and Maltese affairs that he did not seek another showdown between himself and his great enemy. And he revealed no alarm when later told that Napoleon had escaped from Egypt, was back in Paris, and was, thanks to a coup in October 1799, in a stronger position than ever.

Confused loyalties and occasional ill-considered behavior in the company of the Hamiltons and the king and queen of the Two Sicilies ended in July 1800. Sir William Hamilton's recall to Britain forced Nelson, unwilling to leave Emma, to return with them. After an overland journey that took them through Austria and several German states, the trio, by now a source of gossip throughout Europe, arrived in England on 6 November 1800. Nelson's status as a national hero—indeed, as *the* national hero—compelled the Admiralty to make allowances for his erratic behavior in Naples and his affair with Emma. But the Admiralty wanted to occupy his time, hoping it would keep him out of trouble, and promptly returned him to the activity in which he had no equal: combat at sea. On New Year's Day 1801 the Admiralty promoted him to vice admiral of the blue and sent him to join admiral the earl of St. Vincent in the Channel Fleet.

Copenhagen, Amiens, and Unfinished Business

Nelson hoped his new service would bring him and St. Vincent into glorious combat with the French fleet.[107] Yet he suspected and soon learned that he was to head into the Baltic as second in command to Sir Hyde Parker, a very decent but passive and cautious admiral well past his best years. (He suffered from "idleness," Nelson later claimed, which was "the root of all evil.")[108] Parker's mission did not even involve France directly. He was to force Denmark to withdraw from the Armed Neutrality of the North, a confederation recently formed by Russia and the Baltic states to deter the Royal Navy from interfering with their profitable trade with France (ruled by Napoleon as first consul since his coup). If the Danes resisted his diplomacy and threats, Parker was to use all available strength to force a change of mind while keeping the Baltic open to the British shipping that carried supplies vital to the Navy.

At sea and away from Emma Hamilton (to whom he nonetheless wrote several times per day), Nelson finally settled back into naval routine. For the first time since the Battle of the Nile, he began to dedicate himself without distractions to tasks of the highest importance to Britain. He had no special dislike of the Danes, who were Christian, had a strong monarchy, and had no love of republicanism. His sole object, he told St. Vincent, was to help bring the war against the hated French to an "honourable termination." If that meant making an attack on an apparent friend of France, even one with no history of tension with Britain, then he would endeavor to make it truly decisive. Diplomats ("pen and ink men") were useless, he insisted: "A fleet of British ships of war are the best negotiators in Europe; they always speak to be understood and generally gain their point."[109]

Yet Nelson soon realized that Parker had no heart for a fight, even one that served Britain's needs so well. So Nelson wrote him a most energetic letter, hoping to motivate him to strike as soon as possible and with aggression and resolution. Denmark, he wrote with exaggeration, is now

> hostile to us in the greatest possible degree. Therefore, here you are, with almost the safety, certainly the honour of England more entrusted to you, than ever yet fell to the lot of any British Officer. On your decision depends whether our Country shall be degraded in the eyes of Europe, or whether she shall rear her head higher than ever.[110]

Nelson proved successful not only in argument, persuading Parker to adopt his plans, but also in battle, defeating the Danes in the most bitterly contested battle of the entire age of sail. It took place on 2 April 1801 at Copenhagen, which was strongly defended by anchored warships and powerful shore batteries. At one point the situation looked so grim for the British fleet that Parker signaled Nelson to discontinue the battle. "Now, damn me if I do," Nelson said to those around him, ignoring Parker's order. His disobedience and gritty perseverance paid off. In a dreadful slogging match, he virtually destroyed the entire Danish fleet. Then, after Nelson warned the enemy—"the Brothers of Englishmen, the Danes," as he sensitively called them—that greater destruction and bloodshed would follow if they did not surrender, the Danes chose to give up the fight.[111] They recognized defeat and the futility of persistence.

Danish losses were indeed bad, not only in casualties (over 1,700 killed and wounded) but also in vessels, with almost all ships of the line destroyed or taken as British prizes. Nelson's casualty total was also very high (over

940) but, despite his fleet sustaining much damage, he could say with truth that he lost no ships.

Nelson himself went ashore to negotiate an extension to the cease-fire, which later became an armistice. He was most gentle and magnanimous in victory, telling the Danish crown prince that he never counted Danes among his enemies. Quarrels ought not to exist between them. Indeed, he would "consider this the greatest victory he has ever gained, if it may be the cause of a happy reconciliation and union between his most gracious sovereign, and his Majesty the King of Denmark."[112] Nelson's diplomatic efforts were doubtless enhanced by his undisguised admiration for Danish combat courage and tenacity, which provided the Danes with at least some consolation.

Yet in his letters to friends and colleagues, Nelson was smug regarding the widespread praise he subsequently received for his "humanity in saving the town from destruction." He told Emma Hamilton: "Nelson is a warrior, but will not be a butcher,"[113] an ironic comment from a man whose actions in Naples less than two years earlier displayed little humanity. Even so, his liking of the Danes and respect for their valiant performance in battle were genuine. (He never forgot their heroic defense and later often remarked on it.) His attitude in this case represents a sharp contrast to his continued hatred for the French, against whom he would be pitted once more.

Within three weeks of Nelson's return to England from the Baltic, the Admiralty gave him a new task. Perhaps still wishing to keep him out of mischief but certainly wanting his immense popularity to provide reassurance and boost morale across the country,[114] the Admiralty appointed him commander of the naval forces hastily assembled to counter any invasion attempt launched by Napoleon. This attempt seemed likely; Napoleon was busy assembling a flotilla of barges and a large army at Boulogne with the stated intention of invading England. Nelson doubted the army was as strong as claimed, suspecting with typical anti-Gallicism that "some [French] scoundrel emigrant in London" was exaggerating to cause trouble.[115]

Still, he responded to the invasion threat with the expected Nelsonian aggression. He yearned to catch "that Buonaparte" in a duel at sea but considered it unlikely.[116] So believing offense to be the best form of defense, on 15 August 1801 he attempted a raid (his second, following a limited reconnaissance raid two weeks earlier) on the French flotilla in Boulogne.

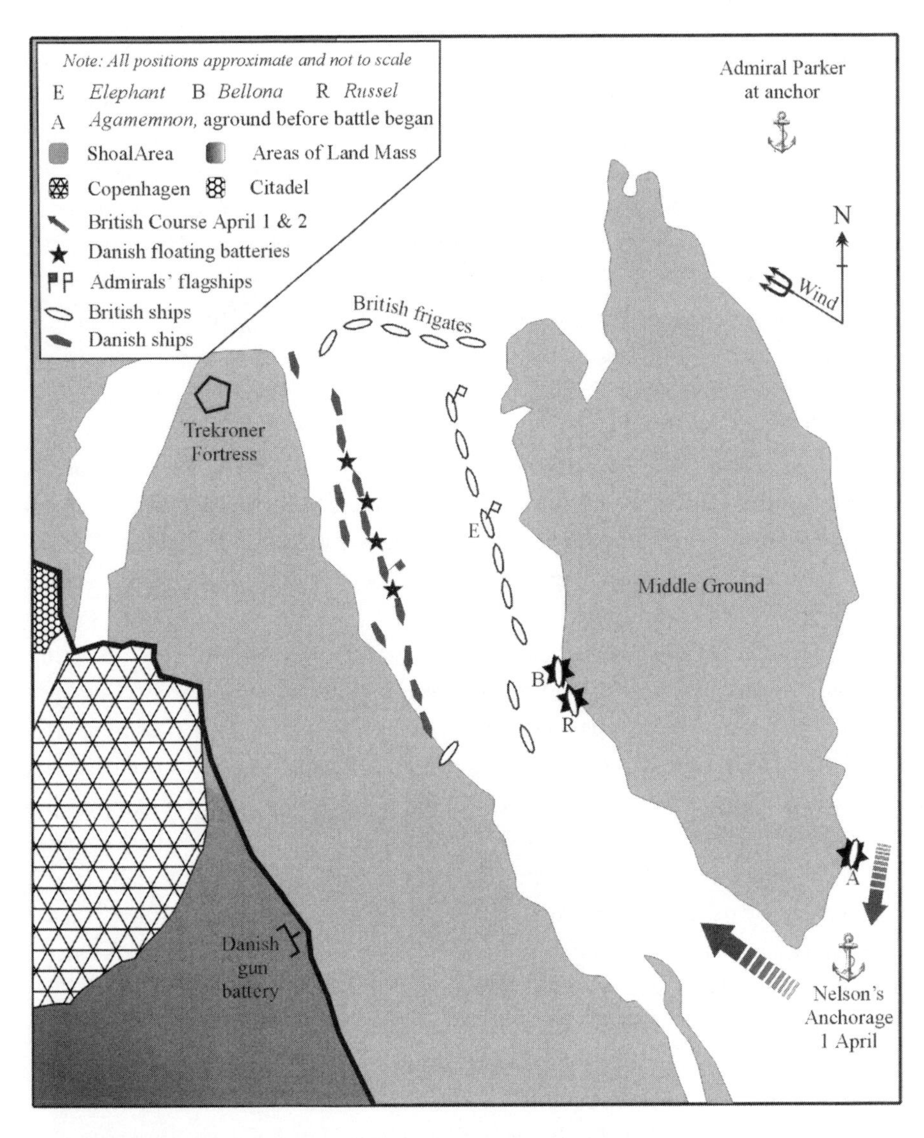

Positions at the Battle of Copenhagen

It proved a tragic failure, his first of any scale since Tenerife four years earlier. Nelson, who as a vice admiral was now too senior to lead these raids himself, was deeply saddened by the loss of 45 men killed or mortally wounded. These included his twenty-two-year-old protégé, Edward Parker, who died after weeks of agony. Another 128 suffered less savage injuries. He promised revenge, telling his squadron that, if the French would only dare to attack, "they will be conducted by his brave followers to a British port [as captured prizes], or sent to the bottom!"[117] He apparently also promised to launch another, retributive attack on Boulogne, which he, vice admiral though he was, would lead in person.[118]

Nelson would have to bide his time. In October 1801 Britain and France entered into peace negotiations, the former needing a respite after the Treaty of Lunéville on 8 February prompted the disintegration of the Second Coalition, and the latter needing time to consolidate its regime at home and reorganize the western German states into a French-protected confederation. After a six-month armistice, the two protagonists signed a formal peace treaty at Amiens on 25 March 1802.

British public support of the armistice and treaty was so widespread it easily swamped the views of those who felt less enthusiastic. Napoleon's reputation soared in many quarters, where he was hailed as the maker of peace.[119] Nelson found this confusing and frustrating. After the negotiations commenced, he told St. Vincent that while the prospect of peace was pleasing, all efforts should still be made "to keep French men and French principles out of our happy country."[120] And unlike St. Vincent, who enthusiastically supported the peace,[121] as did many other senior sea-officers, Nelson remained deeply suspicious of French motives and gave occasional public morale-boosting addresses in which he encouraged a strident anti-French stance.[122] He even privately cursed as "damned scoundrels" the Londoners who excitedly pulled through the streets the carriage of the diplomat carrying the Ratifications of the Preliminaries of Peace.[123] "I am ashamed for my country," Nelson angrily wrote.[124] His bitterness can be contrasted to the general public's euphoria, but his feelings should be viewed in light of his intense grief at young Parker's death after six weeks of agony. Nelson had wept openly at his funeral, having often referred to Parker affectionately as his "son."[125] Consoled by Emma Hamilton, Nelson's pain and sense of loss slowly eased during the next year of peace, but his determination to provide a final end to Napoleon, now First Consul for Life, did not.

Permitted to strike his flag and semiretire, Nelson remained on shore throughout the Peace of Amiens, which proved a recuperative period of relaxation with Emma, whose husband had recently passed away. Yet he kept a close eye on Continental affairs, especially those involving Napoleon, whose professed quest for international cooperation and harmony he refused to take seriously.[126] Indeed, Nelson saw peace with France only in terms of mutual nonaggression—along with many anxious captains hoping not to be retired on half-pay.[127] That is, they saw it as a cease-fire between exhausted belligerents and not as a genuine and potentially lasting détente.[128] "I am the friend of Peace without fearing War," Nelson explained in May 1802, shortly after the formal treaty was signed, "for my politics are to let France know that we will give no insult to her Government, nor will we receive the smallest."[129]

Yet Britain did accept Napoleon's insults—for a year anyway. Napoleon used the Peace of Amiens to expand French power by annexing Piedmont, Elba, and part of Switzerland. He made himself president of the Italian (formerly Cisalpine) Republic, and pressured Holland and Naples into attenuating their trade with Britain. He refused to negotiate a promised trade agreement between France and Britain and became increasingly and publicly hostile over the latter's unwillingness to surrender Malta. Britain's patience finally wore thin, its ambassador being withdrawn from Paris in April 1803 and its fleet returning to active operations without a formal declaration of war. Napoleon responded by occupying Hanover, a family possession of King George III, and by once more commencing the build-up in Boulogne of a powerful invasion fleet, self-righteously claiming that at long last, Britain's "greed and ambitions are out in the open; the mask is down."[130] The fiction of peace dissolved; the reality of war returned.

From this time until Wellington and Blücher defeated the French at Waterloo in 1815, the British public, disappointed to see peace slip away after only a year, displayed considerably more anti-French hostility and unity in support of its war efforts than it had throughout the 1790s.[131] Napoleon himself received principal blame for the collapse of peace and became, consequently, "fully established as a proper object of detestation, and for the next dozen years this feeling was maintained" by the constant burden of war.[132]

Nelson never gloated over his correct assessment, nor rejoiced, as he had on earlier occasions, to learn in May 1803 that the Admiralty was sending him back to sea. Physical weariness and domestic tranquillity with the woman he loved, coupled with a realization that he was already the nation's paramount hero and could attain no greater glory—except perhaps through death in battle—had robbed the forty-four-year-old of some of his earlier passion for war. Thankfully for Britain, his inner feelings and the circumstances of his life had not eroded his resolution to keep Britain safe, his disgust at French rapacity, or his desire to defeat Napoleon at sea. On the contrary, Napoleon's "failure" to honor the Treaty of Amiens had convinced Nelson and his naval peers that the most severe measures were now needed. Indeed, it was Collingwood who claimed that Napoleon possessed "as much villainy as ever disgraced human nature in the person of one man."[133]

Nelson, Britain's favorite son, felt war weary but totally committed to meeting his nation's expectations, which were now great. St. Vincent, serving since 1801 as first lord of the Admiralty, praised Nelson for his "many proofs of transcendent zeal" in the service of king and country and expressed a prayer for the preservation of his "invaluable life" so that he could do "everything that can be achieved by mortal man."[134] St. Vincent then named him commander in chief of the Mediterranean Fleet. This and the Channel command, which went to Admiral the Honorable William Cornwallis, were the largest and most critical Royal Navy operational commands. For a sea officer like Nelson, with no interest in desk jobs in London, this was the pinnacle of success.

Yet it was a tough assignment. The Admiralty instructed Nelson to defend Gibraltar, Malta, and the Kingdom of the Two Sicilies. They also instructed him to blockade the French fleet in Toulon, to bring it to battle should it venture out, and at all costs, to prevent it ever escaping and joining up with the French fleet at Brest. That combined force—the greater part of the French Navy—would doubtless try to gain command of the English Channel and assist Napoleon's invasion of southern England. It must never happen, as Nelson well knew.

Blockade duties were tedious, exhausting, and logistically demanding, for Nelson as well as for his friend Cornwallis, who was blockading Brest during this same period. Nelson, with the harder mission (because of severe supply and maintenance difficulties), conducted it with a maturity, calmness, and patience never before revealed. His craving for glory had given way to a deep and sometimes solemn sense of duty. As one writer has recently commented: "Distant from England and Emma, and free from both ambition and anger, he acquired an emotional stability that had long been lacking."[135] For almost two years, the fatigued and often seasick but always focused Nelson patiently sat outside Toulon. He tried not to keep Admiral Villeneuve's fleet contained in Toulon, as traditional blockade tactics demanded, but tried to lure it out through the pretense of a distant and careless blockade. Once the fleet left Toulon, he reasoned, he could destroy it in battle and thus finally rob Napoleon of the means to interfere with British interests in the Mediterranean or to attack Britain itself. Admiral Nelson could then retire for good, knowing he had done his job.

In anticipation of such a climactic battle, he kept his fleet healthy, well trained, and in a pleasing state of morale. "I believe we are uncommonly well disposed to give the French a thrashing," he told St. Vincent in September 1803, "and we are keen."[136] He still saw his mission in terms of destiny, believing that God had let him survive and win countless battles so that he could deliver a final knockout blow to French overseas ambitions. He was no longer solely or even primarily motivated by revulsion for republicanism and its seemingly inevitable trail of misery and devastation. Even though he still considered republicanism the opposite of his own belief system, and remembered with the sourest distaste the murder, violence, and chaos he had witnessed, he considered that republicanism had been largely swept away by the new and artificial aristocracies and monarchies created by Napoleon and his ruling elite. These were no less destructive or bent on conquest, however, and they and their false and insatiable monarch now served as the focus of his utter contempt.

Like many of his contemporaries, Nelson had come to see Napoleon himself (as the English would later see Hitler) as the greatest hindrance to a lasting peace. The Corsican's ambition seemed limitless; his regard for others' sovereignty nonexistent. Nelson mentioned Napoleon often, yet always disparagingly. He was unable, or unwilling, *ever* to praise him for possessing warfighting talents that—apparently alone in Europe—equaled his own.[137] "Buonaparte's tongue is that of a serpent oiled," he told one Mediterranean ruler in 1804, the year the Corsican became emperor of France. "Nothing shall be wanting on my part to frustrate the designs of this common disturber of the human race."[138] Indeed, Nelson wanted to do far more than merely "frustrate" Napoleon's plans. After the collapse of the Peace of Amiens convinced him that Napoleon could only be reasoned with by military and naval force, he began repeatedly predicting that he would not only defeat, but actually "annihilate," Napoleon's fleet.[139]

On 30 March 1805, however, Napoleon's Toulon fleet (without the emperor himself, who was preparing his invasion fleet along the English Channel) took advantage of bad weather to slip through Nelson's blockade. Without being detected, Villeneuve led the fleet out of the Mediterranean and westward across the Atlantic toward the West Indies. After Nelson's frigates vainly combed the Mediterranean in search of the French, the disconsolate admiral trusted his limited intelligence sources and also sailed for the West Indies. He felt intensely distressed to have lost the French and placed Britain and its interests at risk. One of his marine officers later wrote that the admiral was "said to have behaved like a madman when he found they had escaped him."[140] Anguish filled Nelson's subsequent letters: "My good fortune seems flown away. I cannot get a fair wind, or even a side wind. Dead foul!—Dead foul! . . . I believe this ill luck will go near to kill me."[141] He described his "present anxiety" to his friend Alexander Davison: "I can neither eat, drink or sleep."[142]

His morale collapsed, but his resolve did not. "I cannot forego the desire of getting, if possible, at the Enemy," he wrote to Henry Addington, now Viscount Sidmouth, the recently retired prime minister, "and therefore, I this day steer for the West Indies. My lot seems to have been hard, and the Enemy most fortunate; but it may turn. Patience and perseverance will do much."[143]

He was right. Villeneuve did not remain in the West Indies and promptly sailed back to Europe with the dogged Nelson, steadily closing the gap, in hot pursuit. It was a three-month, seven-thousand-mile round-trip voyage. The French admiral arrived in Cadiz on 20 August 1805, where French and Spanish reinforcements swelled his fleet to thirty-nine ships of the line.

A smaller British squadron hastily attempted a blockade before the Admiralty, having given Nelson brief home leave to recover from nervous and physical exhaustion, sent him back to do what he did best: destroy the enemy and prevent national disaster. He had received no blame for his efforts throughout the previous two years, not even for his failure to catch Villeneuve.

Indeed, his tenacity and total dedication inspired the government, the public, and the Royal Navy alike.[144]

Still exhausted after his long pursuit, Nelson nonetheless responded with typical devotion to duty. He would punish that damned enemy for leading him on such a "pretty dance." He joined his new squadron, enlarged to a fleet of thirty-four ships of the line, and planned his battle to end Napoleon's overseas ambitions once and for all. That battle is so well known that the merest narrative is all that is needed here.

On 19 October 1805, Villeneuve sailed from Cadiz with thirty-three ships of the line, apparently hoping to reenter the Mediterranean. Nelson sent twenty-seven ships of the line to intercept him, catching him off Cape Trafalgar on 21 October. Nelson's remarkable tactics—breaking the enemy line in two places with separate but cooperating squadrons—led to the massive and truly decisive battle against the French and their allies he had hoped to attempt. And it worked marvelously. After five hours the remnants of the combined Franco-Spanish fleet broke off battle, having lost nineteen ships captured or destroyed but having gained not a single success over a British ship and crew.

Britain's stunning triumph was balanced, however, by a loss of the greatest magnitude: an hour after the battle's outbreak, Nelson received a mortal wound from a French sharpshooter. He lived only long enough to learn that he had gained the victory he craved, which at last caged Napoleon in the Continent. He died shortly after whispering: "Now I am satisfied. Thank God, I have done my duty."

Trafalgar did not totally destroy Napoleon's navy. But in the ten years that elapsed before the emperor's final defeat at Waterloo, his Brest fleet and the Toulon fleet that he tried to rebuild proved woefully incapable of causing serious menace anywhere or challenging the Royal Navy's now-confident mastery of the sea. Napoleon's ambitions remained firmly bound on the Continent. Napoleon had many more stunning victories on land. (As it happened, his spectacular victory at Ulm occurred only one day before his great defeat at Trafalgar, allowing him to misrepresent the latter's scale and impact to the French public.) Yet Napoleon could never again threaten the British with invasion, and his fleets could do no other mischief than commerce raiding. Nelson would indeed have been pleased.

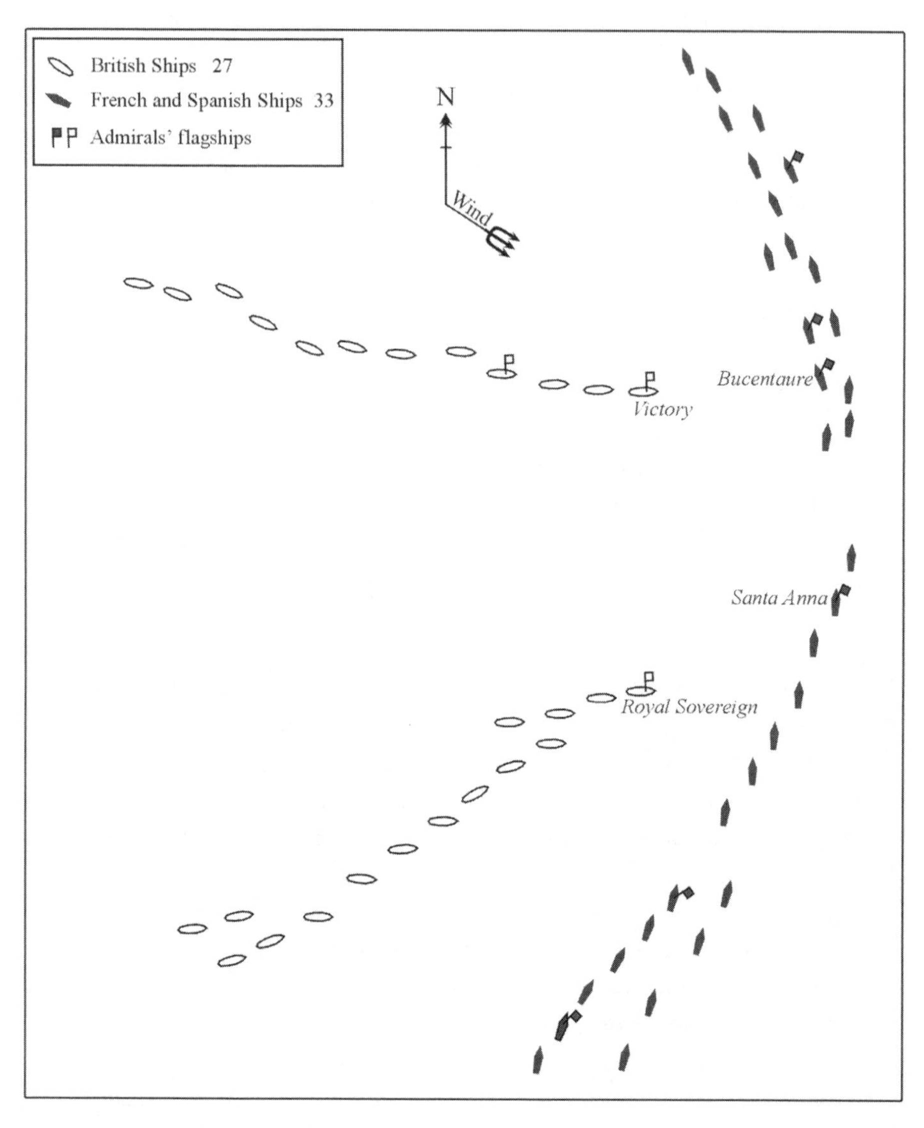

Positions at the Battle of Trafalgar

Conclusions

Horatio Lord Nelson was a remarkably complex man, with passions and emotions seldom disguised and sometimes prevailing over rational decision making. He loved his God, his country, his navy, his wife, and then his mistress, and he craved glory and public adoration; but he also hated his nation's enemies, particularly the French, with equal intensity. As with his other objects of focus, his feelings about his enemies changed over time and were sometimes less fervent, sometimes more so.

The American War of Independence dominated the early years of Nelson's active career; yet, although it caused great angst in the court of King George III, it made little impression on the young seaman. His only known hostility toward the new United States had nothing to do with a diminution of Britain's colonial strength, or even with republican and antimonarchical ideals, but with casual American trade challenges to his overinflated sense of importance and his excessive legalism during his tour of duty in the Leeward Islands. Equally importantly, he then showed no noticeable hostility toward the French, not even for the substantial contribution they made to the American victory over Britain. He certainly enjoyed his own three-month holiday in France.

His hostility toward France increased dramatically, however, after he returned to active service during the French Revolution. At first believing that the fear and disorder within France were essentially matters of domestic politics, he gradually responded to France's expansionism and ideological fervor and the threat they posed to British interests by intensifying his hatred for the rapacious, atheistic republicans. Atrocious French violence following the recapture of Toulon shocked him (and all other British observers), as it did in other places following French military successes. Then, when he saw that French royalists were themselves a burden to Britain—and needed feeding, arming, and protecting—but seemed unwilling to help the cause common to both countries, he became hostile to *all* French.

The First Coalition's eventual failure caused embarrassment and frustrations and strengthened Nelson's anti-Gallic feelings, which became even more intense after Napoleon's series of victories against the Austrians earned him fame and glory—but robbed it from Britain (and from Nelson). Successes and losses against the Spanish, for whom Nelson had no particular hatred, did little to distract his mind from the pursuit of revenge against France and its glorious general, Napoleon. Nelson obtained his revenge at the Battle of the Nile.

Yet Napoleon was humbled at the Nile, not broken, and his eventual return to France and greater power, achieved through a coup, left Nelson with unfinished business. He would finish it, but not for six years. In the meantime he took his anger out on Neapolitan rebels, whom he hated for their French-influenced brutality and disloyalty, and waged war on Napoleon's allies, particularly the Danes, of whom he had no inherent dislike or feeling of passionate enmity. He thus hoped to serve his country and punish Napoleon—"whatever name he may choose to call himself; General, Consul, or Emperor," as he later wrote—through indirect means.[145]

Napoleon's purported blame for the collapse of the Peace of Amiens convinced Nelson that the emperor was so capricious and insatiable that only the "annihilation" of his fleets would cage him in the Continent. The huge Battle of Trafalgar in October 1805 finally gave Nelson the naval victory over Napoleon and France he had long craved, even though it robbed him of life at the very moment of triumph. It did not spell defeat for the French emperor, much as it wounded his pride and ruined many of his ambitions, but it did bring safety, security, and vast prestige to Britain and allowed the Royal Navy to control the sea (which it did for the next century).

Hatred is ordinarily a miserable emotion; but in Nelson's case, it inflamed his patriotism and kept it burning—sometimes less brightly, other times more so—throughout two of Britain's darkest decades. Based as it was on a loathing of what and whom he considered evil and hurtful and not on any pathological desire to destroy or dominate others, Nelson's hatred curiously enriched his complex personality. When coupled with the many positive aspects of his psyche—intuition, love, passion, tenacity, courage, and audacity—his hatred and aggression made him a fearsome warrior: good to have on one's side, but not to face as an enemy.

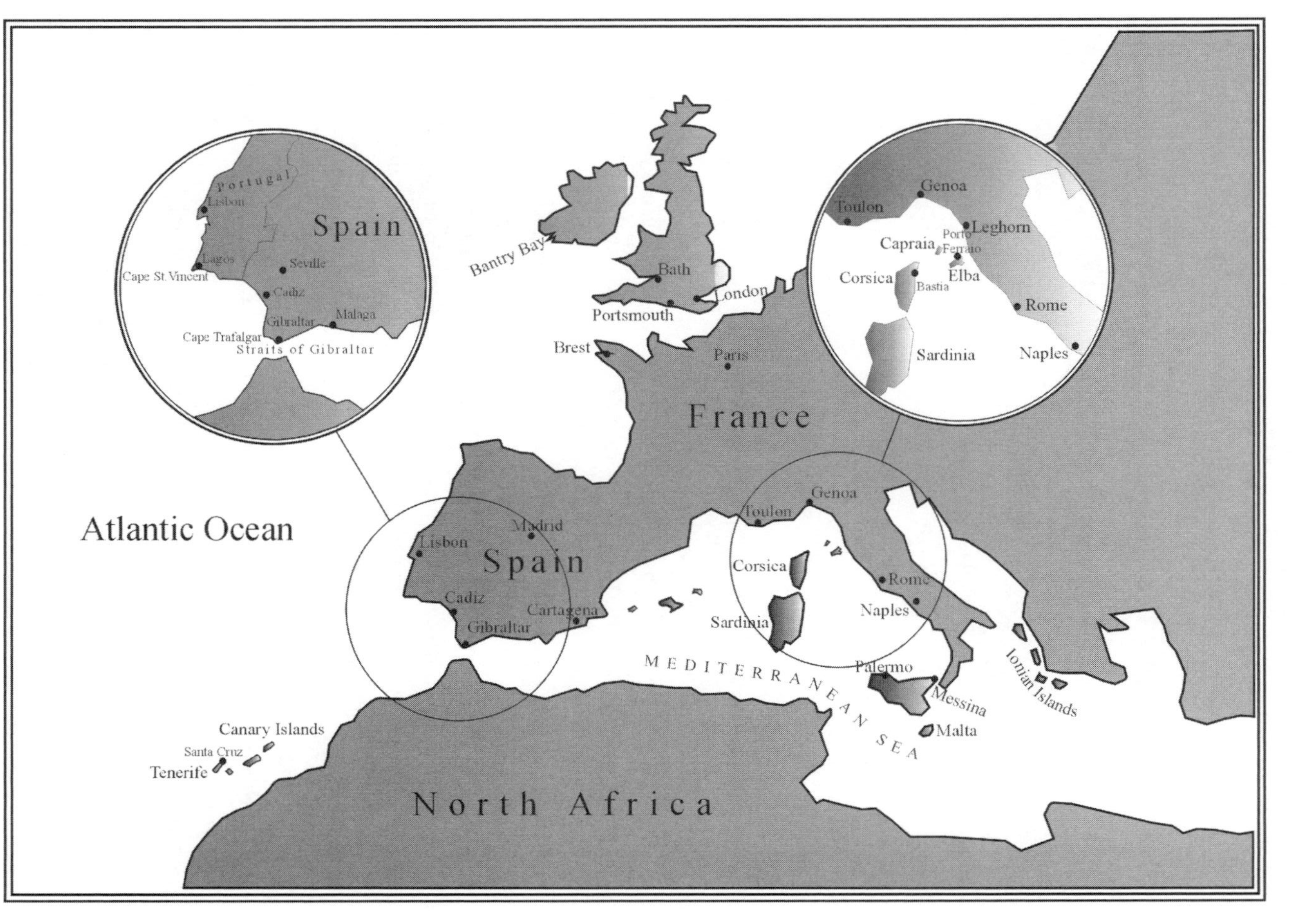

Europe, the Mediterranean, and North Africa

2

The Admiral's Spiritual Beliefs

NELSON IS famous not only for his great victories at the Nile, Copenhagen, and Trafalgar, and for the martial genius and battle fervor that produced them, but also for his unashamed, torrid, and adulterous love affair with Emma Lady Hamilton, the wife of a British diplomat. The blending of heroism and romantic love is not uncommon, of course, and appears repeatedly in both factual and fictional representations of war and warriors. Yet Nelson was truly complex, with a constant paradoxical balancing of selfish and selfless motives and actions, caused by an obsessive craving for fame and glory, which doubtless increased his combat valor but also prompted some vain and foolish behavior. And it was also caused by an unswerving loyalty to both king and country, which certainly motivated him to overcome serious physical disabilities so that he could continue to contribute to his nation's well-being.

Most unusual, at least given Nelson's unbridled ferocity, extreme vanity, and seemingly remorseless adultery, was his lifelong adoration of the Christian God and the gospel of His son, Jesus Christ. Yet the origins of Nelson's profound religiosity, which went well beyond the piety and church attendance common to Anglicanism of that era, are easy to explain: he was the grandson, son, and brother of Anglican ministers and experienced a boyhood in which scriptures, prayer, and church services played significant roles. Additionally, the dominant figure of his early career was Capt. William Locker, a paternal officer and observant Christian, whose righteous conduct and wise counsel clearly reinforced Nelson's earliest theological framework. Nelson adored Locker, who also taught him invaluable tactical lessons, and Nelson's many letters to him, spanning over twenty years until Locker's death in 1800, often expressed rich Christian imagery, prayers, and blessings.

Right up until his premature death in 1805, Nelson did many things that seem hard to reconcile with Christian principles, as he was himself painfully aware, yet he never ceased believing in God's sovereignty, omnipotence, and omniscience. Indeed, he routinely expressed his Christian beliefs in dispatches, letters, and conversations, as well as in divine services and other church ceremonies. His own last diary entry, written shortly before his death at Trafalgar, contains these beautiful words:

> May the Great God, whom I worship, grant to my Country, and for the benefit of Europe in general, a great and glorious Victory; and may no misconduct in any one tarnish it; and may humanity after Victory be the predominant feature in the British Fleet. For myself, individually, I commit my life to Him who made me, and may his blessing light upon my endeavours for serving my Country faithfully. To Him I resign myself and the just cause which is entrusted to me to defend. Amen. Amen. Amen.[1]

This chapter attempts to describe the great admiral's religiosity and to explain how he reconciled his excessive pride, his ruthlessness during battle, and his adulterous union with Emma Hamilton with his lifetime devotion to the gospel of Jesus Christ. By retracing Nelson's footsteps with a particular focus on his toughest moments and greatest battles (which were not necessarily the same events), the chapter also tries to make sense of the extent to which Nelson's religious views shaped the ways he perceived and practiced naval warfare, and commanded and led his subordinates.

Genesis

In 1799 Nelson wrote a short autobiographical sketch of his life.[2] Matter-of-fact and lacking the personal and often intimate passages that make his voluminous correspondence with friends and colleagues such compelling reading, this short sketch describes his birth and childhood in a mere two sentences.

These tell us only that he was born in the parsonage house at Burnham Thorpe, in Norfolk, to the Reverend Edmund Nelson, the local rector, and his wife Catherine, daughter of Dr. Maurice Suckling, prebendary of Westminster. This information, although naturally correct, fails to convey the strong commitment of his family, on his father's and mother's sides, to the Anglican faith. Both his grandfathers were Anglican clergymen, and two of his great-uncles, eight cousins, and two of his brothers also took holy orders.

Nelson's mother died when he was nine years old, leaving the rector to serve as a double parent to eight children, the youngest only ten months old. Despite his lugubrious and self-deprecating nature, Edmund Nelson proved a dedicated and loving father, who, through his own example and instructions, tried (certainly successfully in the future admiral's case) to instill in his children a love of knowledge, strong familial bonds, a strict moral code, and an awareness of God's omnipotence.

Extant sources reveal few glimpses of the childhood of Horatio and his siblings.[3] We know they interacted harmoniously and with no dysfunction (to use today's parlance), and can deduce from later evidence that they must have studied the classics and at least some of Shakespeare's works. Yet we don't know, for instance, whether they conducted extra bible study between church services, whether they prayed together before meals or in the evenings, or whether they received encouragement to read any of the religious tracts and devotionals that were popular at that time. One can only presume that the Nelson children, regardless of specific childhood activities, received an upbringing typical of the offspring of clergy. That was certainly the view of the Reverend Alexander John Scott, who knew the admiral for many years and served as his sea chaplain from 1801 to 1805. Scott recalled that Lord Nelson was "a thorough clergyman's son" and that, for example, even when he had risen to vice admiral, he doubtless "never went to bed or got up, without kneeling down to say his prayers."[4] Nelson himself admitted to this prayer pattern: "When I lay me down to sleep," he wrote in his journal after an engagement with French warships in 1793, "I recommend myself to the care of Almighty God. When [I] awake I give myself up to His direction."[5]

In any event, Nelson left his family in 1771 for a naval career at what today seems a dreadfully premature age—twelve years—even though this age was then quite normal for the undertaking. The young boy passed into the guardianship of his uncle, Maurice Suckling, a respected captain, and then into the care of a series of other merchant and naval captains. The most important of these, at least in terms of his influence on the young man, was Captain Locker, "who loved Nelson like one of his own children."[6] Locker had once served directly and closely under the legendary Adm. Sir Edward Hawke, whose tactical flexibility brought outstanding results in battle. Locker was thus able to give Nelson a marvelous schooling in naval tactics. "Lay a Frenchman close and you will beat him," Locker used to say, referring to the superior tactics, zeal, training, and rate of fire of British crews.[7] Almost two decades later, when Locker was lieutenant governor of Greenwich Hospital, Nelson attributed much of his own great success to Locker's influence: "I have been your scholar," he wrote to the old man, "it is you who taught me to board a Frenchman . . . and my only merit in my profession is being a good scholar."[8] Their friendship would never end, he added, except when death took one of them away.

Locker's influence did not relate only to tactics. The well-read, sharp-witted captain was a paternal officer and faithful Christian, whose character, conduct, and counsel provided Nelson with a leadership model that he would emulate always thereafter, a model that reinforced his belief that valiant warriors could also be strong Christians. When Locker died in December 1800, the distraught Nelson told his "righteous" old mentor's son John that "the greatest consolation to us . . . is that he left a character for honour and honesty which none can surpass, and very, very few attain."[9] Nelson would "revere" Locker's memory, he told an acquaintance nine months later, to the last moment of his own life.[10]

Indeed, Nelson had grown to love Locker as he loved few others outside his immediate family. His many letters to him, spanning over twenty years until Locker's death in 1800, reveal that he felt no unease about including Christian imagery, prayers, and blessings. When Locker fell gravely ill in 1777, for example, the anxious young Nelson wrote: "I hope God Almighty will be pleased to spare your life. . . . all the services I can render to your family, you may be assured shall be done, and shall never end but with my life." He closed his letter with a heartfelt blessing: "May God Almighty of his great goodness keep, bless and preserve you, and your family, is the most fervent prayer of your faithful servant, Horatio Nelson."[11] Throughout the next twenty-three years, Nelson would write scores of letters to Locker, telling him candidly how he felt about various events and circumstances and often mentioning how he saw God's hand, which Nelson frequently described (not uncommonly for the era) as Providence, at work to protect him and his crew.[12]

After an engagement with a French squadron off Genoa in March 1795, in which Nelson's outstanding tactical skills permitted him to inflict significant damage to an enemy vessel, despite its superior strength and firepower, he told Locker: "*Providence,* in a most miraculous manner, preserved my poor brave fellows. . . . one hundred and ten of the enemy were killed and wounded on that day, and only seven of ours wounded."[13] Interestingly, only a few days earlier he had told his wife, Fanny, in anticipation of this very engagement: "Whatever may be my fate . . . the lives of all are in the hands of Him who knows best whether to preserve mine or no, and to His will do I resign myself."[14] Thus, in Nelson's mind, Providence referred to the Almighty's protective intervention in the actions of His servants.

His best friend Locker was not alone in enjoying warm and unaffected correspondence with the great sea officer. Most recipients of Nelson's letters and dispatches—including many who were close to him in other ways but perhaps not in matters of faith—received relatively few unveiled statements of Christian belief and devotion. Yet his letters to his father, wife, and paramour expressed matters of the heart and the spirit with obvious comfort and sincerity. Actually, right up until Nelson became entangled with Emma Hamilton in 1798, his most candid letters went to his closest confidant: Fanny, whom he courted for eighteen months before marrying on 11 March 1787.

Even his earliest letters to Fanny contain explicit statements of Christian faith. On 11 September 1785, the young captain, then rigidly enforcing the Navigation Acts in the Leeward Islands, tried to console Fanny after the death of Sarah Herbert, her close relative. "But this believe from my heart," he wrote, "that I readily partake of all the sorrows you undergo. . . .[but] call religion to your aid, and it will convince you, that her [Sarah's] conduct in this world was such as will ensure everlasting happiness in that which is to come."[15] Six months later, with marriage firmly in his mind, he penned a beautiful letter to Fanny: "Separated from my dearest what pleasure can I feel? None! Be assured all my happiness is centred with thee, and where thou art not there I am not happy."[16] Their courtship was God's will, he added: "I daily thank God who ordained that I should be attached to you. He has I firmly believe intended it as a blessing to me, and I am well convinced you will not disappoint his beneficent intentions." Although he lacked wealth, he trusted that "the Almighty who brings us together will, I doubt not, take ample care of us, and prosper all our undertakings." He even saw his future responsibility for Fanny's son Josiah (whose father had died a few years earlier) as a gift from God: "Let me repeat that my dear Josiah shall ever be considered by me as one of my own. That Omnipotent Being who sees and knows what passes in all hearts knows what I have written to be my undisguised sentiments towards the little fellow."

From these two letters and the two already mentioned that describe the naval engagement off Genoa (and there are scores more that express the same Christian principles with no inconsistency), one can gain a clear and reliable understanding of Nelson's theological beliefs. He was a typical rural and then naval Anglican of his era and upbringing who remained largely untouched by the rationalizing influences of the Enlightenment. Indeed, throughout his voluminous correspondence, he makes no mention of the philosophers David Hume and John Locke or the latter's great ideological descendant Thomas Paine. And he appears never to have read any of their publications.[17] On the other hand, he once favorably quoted an aphorism from Edmund Burke, which is consistent with the view that Nelson rejected reactionary thought and embraced that which upheld conservative values and the established social and political order.[18]

He devoted himself to a creator with whom he could experience an active and deeply personal relationship. He clearly believed in God's omnipotence —that He was all-powerful and that no other force in the spiritual or natural realm could interfere with His own designs for peoples and individuals. He also believed that God was omniscient, knowing all and capable of seeing into the hearts and minds of His creations, who could hide nothing from Him but could communicate their desires, needs, and fears to Him through prayer. With both omnipotence and omniscience, God exercised divine government in order to manage nature and guide history in keeping with His grand plan for mankind. This divine management often necessitated direct intervention—the guiding hand of Providence—in human affairs.

Nelson saw no tension or contradiction between God's lordship and the free will and consequent responsibilities of individuals for their actions. He clearly believed in sin and its consequences, and he feared God's wrath. Yet he also understood that devotion to scripture (the *viva vox*, or living word), faith in Christ, and repentance of sin would restore the sinner, now made righteous, to a right and proper relationship with God. That relationship would guarantee the righteous person salvation and eternal life in God's presence.

The sea officer adhered to these tenets of faith with greater fervor than most of his famous contemporaries, at least judging by their actions and correspondence. He believed in the tenets unquestioningly. Many modern readers might assume that Nelson's beliefs are incongruous with a warrior's voluntary contribution to battles of great destruction, terror, and chaos. War certainly seems a most ungodly activity. As Gen. William Tecumseh Sherman told an audience in August 1880: "There is many a boy here today who looks on war as all glory, but, boys, it is all hell." General Sherman's famous speech sums up what most people who have lived through the privations and horrors of war—service personnel and civilians alike—actually believe about their experiences. War is frightening and destructive and tends to bring out the worst in people: hatred, cowardice, violence, brutality. It *is* hell.

Yet, paradoxically, war also brings out the very best human traits: love of country and kin, self-sacrifice, courage, and comradeship. In many cases the ever-present specter of pain and death also raises interest in spiritual issues and creates a greater reliance on God's grace, comfort, and protection. Countless unknown warriors, and even many of history's greatest combat leaders—with Gen. Robert E. Lee doubtless at the forefront—fall into this category. Devout and God-fearing, they have seen true value in seeking God's counsel and blessings before, during, and after battle.

The eighteenth-century Royal Navy had several such men, including Vice Adm. Edward Vernon, who wrote to his wife after a stunning victory in March 1741: "My Dear—After the glorious success it has pleased Almighty God so wonderfully to favor us with, Whose manifold mercies I hope I shall never be unmindful of, I cannot omit laying hold of the opportunity of an express I am sending home to acquaint you of the joyful news."[19] Nelson was also such a man, perhaps one of the most devout British fighting admirals of his generation, as Vernon had been of his.

As did many of his compatriots, Nelson went beyond the instructions laid out in the Articles of War, which required officers to observe Church of England regulations and make sure ships' officers and men did likewise. Trusting in God's will and blessings as well as in his fleets' abilities, Nelson remained passionately committed to the justice of his cause and devoted all of his energy, instincts, and acquired skills to attain victory over his enemies. But he realized that he could himself achieve only so much. He consequently stressed the importance of prayer, asking God not only for protection and success, but also for generosity and humility if the Lord should give the British victory—and peace and comfort if He should withhold it. One of Nelson's prayers uttered off Cape Trafalgar even asked God for a submissive heart should He choose to take him from this world. Its prescient author died soon after, but with dignity, humility, and a lack of fear. He accepted the Lord's will and looked forward to being in His presence.

Human Weaknesses

Nelson was first and foremost a warrior. Thus, the relationship between his theological views and his apparently contradictory misbehavior—at least vis-à-vis his extreme craving for fame and his adultery—distracts attention from an analysis of what seems more important, the impact of his religious views on the activity he devoted his entire marvelous career to: defeating the enemies of his king and country. Yet his misbehavior or hypocrisy needs explaining, if only to demonstrate that his great vanity and his love for Emma Hamilton also directly impacted upon his warfighting goals and leadership style.

Nelson's extreme vanity and craving for fame seem ignoble when contrasted to his many fine character attributes, and appear incompatible with the Christian concept of humility. After defeating the Danes at Copenhagen in 1801, for instance, Nelson had the arrogance to send Captain Sneedorf, commandant of the Danish Naval Academy, and Adj. Gen. Hans Lindholm, aide-de-camp to the Danish crown prince, copies of a short account of his illustrious life.[20] Young Danes might find it inspiring and benefit from following his example, Nelson said.

The duke of Wellington, that other great British hero of the Napoleonic period, found Nelson's vanity repellent, at least for a short time. In September 1805 he happened to meet Nelson in a waiting room in the colonial office. Although the admiral was instantly recognizable because of his distinctive appearance and missing arm, he did not recognize the soldier, who had yet to earn his greatest laurels. "He could not know who I was," Wellington recalled, "but he entered at once into a conversation with me, if I can call it conversation, for it was almost all on his side and all about himself, and in, really, a style so vain and silly as to surprise and almost disgust me."[21] After eventually learning Wellington's identity, Nelson underwent a "sudden and complete metamorphosis" and began to converse "like an officer and a statesman," his earlier "charlatan style" having completely vanished. Indeed, Nelson and Wellington talked for a further forty minutes or so, leaving the latter to recall that he had probably never had a more stimulating conversation. Nelson was, he concluded, "in different circumstances, two quite different men."

It must be remembered that Wellington was what we would today call a *cold fish,* who disliked displays of emotion, detested flamboyance, and even harshly rebuked ordinary people who wanted to shake his hand or convey their praise.[22] Vain and silly Nelson surely was (and on this occasion he seems to have outdone himself), but his excitable, emotional nature would have been totally incomprehensible to the cold and aloof Wellington under any circumstances. Thank goodness Wellington never witnessed Nelson's adrenaline-charged fervor during combat, or his deathbed request to his beloved Hardy shortly before he left this world. "Kiss me, Hardy," he said. We can only guess what Wellington would have made and later said of the dying admiral's desire for that one last act of physical warmth from someone he loved as a brother.

Nelson had always been vain, self-congratulatory, and desperate for public acclamation. He got his first taste of the latter in March 1795, after naval actions off Genoa earned him praise from the friendly central-Mediterranean states. "I am so covered with laurels that you will hardly find my little face," he conceitedly told his wife. "At one period I am the 'dear Nelson,' 'the amiable Nelson,' 'the fiery Nelson.'"[23] This taste of fame doubtless removed the bitterness of being inadequately praised by Lord Hood for Nelson's vigorous contribution to Bastia's capture nine months earlier.[24]

As it happened, Nelson would never again let his deeds pass unnoticed. After the Battle of Cape St. Vincent on 14 February 1797, he felt annoyed that his own decisive role in the stunning victory had been marginalized. His admiral, Sir John Jervis (soon to be named the earl of St. Vincent), had personally thanked and praised him—profusely, in fact—but Jervis's dispatches to the Admiralty did not single out Nelson for the recognition he thought he deserved. Considering this entirely inappropriate, and refusing to allow his greatest combat achievement to go unnoticed in England, Nelson drafted a series of letters to friends, captioned "A Few Remarks Relative to Myself in the *Captain* in Which My Pendant Was Flying on the Most Glorious Valentine's Day, 1797."[25] In these self-aggrandizing "few words," Nelson explained how he had boldly, on his own initiative, pulled his ship out of the British line to prevent Spanish vessels escaping, heroically engaged several ships simultaneously, and then captured two by boarding the first from his own ship and crossing over the first to board the second. Jervis's victory, he implied, largely depended on Nelson's unique contribution to the battle. His intention in writing this record of his heroic exploits was clearly so that his friends, some of them in influential places, would ensure that these accomplishments became widely known, hopefully through publication in newspapers or the *Naval Chronicle*. This happened. His reworded account appeared in the *Times* and the *Sun* (and eventually in the *Naval Chronicle*) and his reputation with the public and the press soared.[26]

His reputation reached great heights, as did his self-admiration, after his brilliant defeat of the French fleet at Aboukir Bay in Egypt on 1 August 1798. That victory had followed a long pursuit of the French fleet, during which Nelson had told his wife Fanny: "Glory is my object and that alone."[27] Even after his marvelous victory at Aboukir saw him showered in adulation—prompting him to boast repeatedly to Fanny about his rapturous reception in the Kingdom of the Two Sicilies—his craving for glory remained intense. In February 1800 when he was already England's most popular hero, he told Emma, his mistress: "My greatest happiness is to serve my gracious King and Country & I am envious only of glory."[28] Closely paraphrasing Shakespeare's *Henry V*, he added: "For if it be a sin to covet glory, I am the most offending soul alive."

Wanting to reflect his glory at every opportunity, Nelson loved to be seen in his immaculate dress uniform, complete with a glittering collection of medals and decorations: not only the Star of the Order of the Bath and his battle medals, but also his famous *chelengk*, a spray of diamonds presented to Nelson by the sultan of Turkey and worn by the admiral in his naval hat. This dramatic, iconic appearance proved a hit with the British public, who adored Nelson and flocked around him whenever he ventured out in society.[29] Many of his colleagues, on the other hand, thought he looked vain and silly.[30] Lt. Gen. Sir John Moore, the famous British soldier, saw Nelson in Naples in

July 1800, "covered with stars, ribbons, and medals," and noted in his diary that he looked more like an opera star than a great warrior.[31] "It is really melancholy," he added, "to see a brave and good man, who has deserved well of his country, cutting so pitiful a figure." Even some of Nelson's closest confidants thought this. "Poor man!" the earl of St. Vincent wrote after one meeting with Nelson, "He is devoured with vanity, weakness and folly; was strung with ribbons, medals, etc., yet pretended that he wished to avoid the honour and ceremonies he everywhere met with on the road."[32] Such false modesty was absurd in the foolish admiral.

To be fair to Nelson (and St. Vincent was not entirely fair), he did not adopt this flashy appearance, resplendent in dress uniform and glittering medals, only because of his desire to be seen and received as a hero for his ego's sake alone. Vain he surely was, but he was also acutely aware that, for most of the revolutionary and Napoleonic periods, Britain was hard-pressed, often fighting alone or with unsuccessful allies, and in need of something or someone to boost morale. In short, Britain needed heroes—and those who would act and look like heroes. Nelson realized that the eyes of his nation were upon him and he feared disappointing the public or appearing as anything else than what they wanted to see: a conquering hero. In this way, then, his flashy appearance and obvious self-confidence were not only an unintended consequence of his great vanity, and perhaps a means of drawing attention from his frail and disfigured frame, but also a conscious attempt to meet his nation's expectations and help morale. In this he undoubtedly proved successful.

Moreover, this vain man was also England's greatest warrior, with combat successes far outstripping St. Vincent's and any other Royal Navy commander's. And, although few scholars have ever made the connection, Nelson's vanity and craving for honor and renown doubtless contributed significantly, and in a very positive way, to his warfighting successes. In particular, the pursuit of ever greater glory drove him to commit acts of remarkable and inspiring heroism, including the boarding of enemy ships and sword-in-hand attacks on other enemies even after he had reached flag rank. It also prompted him to plan operations of unequaled audacity, such as his bold but very dangerous two-squadron attack to break the combined Franco-Spanish line off Cape Trafalgar in 1805.

Further, unlike most of his peers, Nelson never set as a principal objective the prize money that captains and crews could make by capturing enemy ships and transferring them into Royal Navy custody. In many engagements of the era, a squadron would engage a similar enemy force, not with the aim of taking or destroying as many vessels as possible through firepower and tactical maneuver but, rather, with the hope of separating a few ships from their parent squadron and taking them as prizes, even if the rest escaped. This was considered a very good result. *Not by Nelson!*

Vain and driven by his sense of destiny, he wanted the honor and fame that inevitably followed successful great clashes of skill, will, and firepower. He cared far less for incidental prize money (although he cared deeply that his crews always received whatever they were owed). Honor, he told his wife in June 1794, while fighting on land in Corsica, is "far above the consideration of wealth." Even though he was unlikely to claim prize money while there, he said, his actions would bring honor.[33] He further told her in another letter from the period: "I trust my name will stand on record when the money makers will be forgot."[34] And in 1803 he told Sir Evan Nepean, secretary of the Admiralty: "I look not for prize money, which although it would be most acceptable, I despise compared to fighting and gaining a victory over the French fleet."[35]

At sea, Nelson certainly measured success in different terms than most of his peers. Uninterested in acquiring wealth at the expense of losing chances for greater glory, he refused to let any opportunities for beating the enemy slip through his fingers and always aimed for total victory (which meant the enemy fleet's or squadron's reduction to irrelevance). His actions during the Battle of the Nile on 1 August 1798 are a case in point. Naval battles never occurred at night, and the French commander, Admiral Brueys, assumed that any rational foe would wait until morning to launch an attack.

Nelson was no ordinary commander, however. He had an unusually quick and agile mind; and unwilling to risk losing his chance of gaining the tremendous glory that seemed within his grasp—by letting the French ships slip out under cover of darkness or reinforce their defenses before morning—he quickly formed an audacious, aggressive, and perilous plan for an immediate attack that would bring the decisive result he craved. As already noted, his gamble paid off handsomely.

The great man's pride and obsessive craving for fame had not always brought such dazzling results. Puffed up with self-importance after his decisive and heroic contribution to Jervis's victory at the Battle of Cape St. Vincent back in February 1797, Nelson soon initiated a bold campaign that had dire consequences. After hearing that a Spanish treasure ship had put into Santa Cruz de Tenerife in the Canary Islands, he asked the earl of St. Vincent for permission to attempt to "cut out" the ship from this strongly fortified harbor. It would be relatively easy, he argued. The earl approved the idea, giving Nelson four ships of the line, three frigates, and a cutter.[36]

Nelson's plans went wrong from the start. High winds and adverse currents prevented the assailing force from reaching their destinations during darkness, and the Spanish defenders, once alerted, protected their harbor so stoutly that the landing parties had to withdraw.[37] This was too great a blow to Nelson's and his captains' egos. Ignoring his own best instincts, and even common sense, he responded to the unanimous will of a war council by violating a cardinal tenet of warfare: he chose to reinforce failure. That is, he chose to repeat what had already proven costly and unsuccessful. As he later wrote: "*My pride suffered;* and although I felt the second attack a forlorn hope, yet the honour of my country [really his own honor] called for the attack, and that I should command it."[38] The valiant but doomed and senseless attempt cost him many casualties and almost claimed his own life. He left Tenerife without his right arm.

Nelson accepted the failure with great sadness but honest self-criticism. He knew he was responsible and made no attempt to blame others. Indeed, his capacity for self-censure, evident on this and many other occasions, almost mitigates against his overweening egotism. He cursed his pride for causing such bad judgment. He also recognized God's hand in the events at Tenerife. Perhaps because his pride had been too great, Providence had intervened, decreeing that his actions would bear no fruit.[39] Yet he took comfort that God had not abandoned him. He told Fanny that her son Josiah (his adopted son and a junior sea officer), "under God's providence, was principally instrumental in saving my life."[40]

This was not the only time Nelson felt that God was chastening him for his excessive pride. After his flagship *Vanguard* was dismasted and almost sunk in a raging storm in May 1798, Nelson wrote to his wife: "I ought not to call what has happened to the *Vanguard* by the cold name of accident; I believe firmly that it was the Almighty's goodness, to check my consummate vanity. I hope it has made me a better officer, as I feel confident it has made me a better man. I kiss with all humility the rod [of God]." He regretted that he was such "a proud conceited man."[41] With remarkable frankness, he told the earl of St. Vincent the same thing: "My pride was too great for [any] man; but I trust my friends will think that I bore my chastisement like a man. It has pleased God to assist us with his favor, and here I am again off Toulon."[42]

It seems a pity, then, given Nelson's sincere hopes for greater and lasting humility, that he would soon find his pride swollen almost to bursting point after his successes at the Nile and at Copenhagen. He would in fact wrestle with pride for the rest of his life, acutely aware of its displeasure to God (even if he never fully recognized its negative impact on colleagues), but seldom able to bring it under control.

Adultery

His affair with Emma Hamilton hardly helped. She overwhelmed him with adoration, gushing about his greatness to anyone who would listen and transforming the house she shared with her husband, Sir William Hamilton, and later the house she shared with Nelson, into virtual shrines to his martial glory.[43] Scholars of the admiral, and even his many admirers, see his adultery with Emma and his harsh treatment of Fanny as particularly poor behavior and perhaps the greatest stain in his copybook (his maltreatment of Neapolitan Jacobins in 1799 being the other contender for that honor). Certainly from the vantage point of mainstream Christianity, including the Anglicanism of Nelson's era, his adultery was sinful; he committed it in conscious violation of God's commandments.

Nelson knew the scriptures and church teachings well enough to realize that he was not meeting God's expectations. Yet he was so consumed by love for Emma that he was incapable of breaking free. This is not to say that he never loved his wife Fanny. On the contrary, it is clear from the beautiful letters previously quoted that he had grown to love Fanny before their marriage in March 1787 and that he believed God had Himself ordained their union. There is plenty of additional evidence. During the previous summer, for instance, when his naval duties often took him away, he wrote to her: "My heart yearns to you, it is with you, my mind dwells upon nought else but you.

Absent from you, I feel no pleasure; it is you, my dearest Fanny, who are everything to me."[44] After Nelson lost his arm in 1797, the attentive Fanny lovingly nursed him back to health and with patience and tenderness helped him adjust to life as an amputee. He responded with sincere love and gratitude, requesting that Fanny—who was, he said, "beautiful, accomplished" with "angelic tenderness . . . beyond imagination"—be permitted to join him at semi-official functions from which she would ordinarily be excluded.[45] At such functions "his attentions to her," as one observer noted, "were those of a lover."[46]

Despite this and other similar comments, many writers now argue that (to use today's expression) Nelson may have loved Fanny but was not *in love* with her. This argument may contain some truth; Fanny found her husband's periods of sea service increasingly hard to cope with, and to his frequent annoyance, she filled her letters with too many expressions of depression and too few statements of support and praise for her husband's actions. The enthusiasm, warmth, and intimacy of their early years began to wane, and the six months or so of closeness following his amputation may have been too little and too late to restore and maintain romantic love once he went to sea again in 1798. Yet matters of the heart are so deeply personal that even this argument—that they gradually drifted apart until their relationship rested more on formality and responsibility than on love—is speculative at best.

Several things are evident, however: Nelson did fall in love with Emma Hamilton while he was helping to manage central Mediterranean affairs from Naples in 1798 and Palermo in 1799, and the beautiful, vivacious Emma reciprocated his love with equal ardor. He initially felt torn between Fanny and Emma (around this time he was still claiming that his happiest day "was that on which I married Lady Nelson");[47] It also appears that Sir William Hamilton was not fully aware of, or not too uncomfortable with, his wife's increasingly intimate relationship with the admiral, whom he considered a best friend.

Sir William's unconcerned attitude and the reality that Fanny was in distant England and had, in any event, only spent six months with Nelson throughout the entire previous five years do not change the fact that Nelson's affair with Sir William's wife was adulterous—which caused the Christian admiral great emotional and spiritual agony, especially after his return to England with the Hamiltons in November 1800 brought the issue to a head. Something had to be done; a choice had to be made.

After two months of intense soul-searching, Nelson made his choice: in favor of Emma. He formally parted from Fanny on 13 January 1801, never to see her again, and instructed her by letter two months later that although he sincerely wished her well, he wanted no further contact.[48] She would receive a handsome allowance and be taken care of in his will. In this way, it seems, he considered himself as free of his marital vows of love and fidelity as if Fanny no longer existed. Fanny was deeply wounded but never imposed herself in her husband's life, preferring instead to live in quiet dignity and to care for Nelson's elderly father, who found his son's marital problems distressing.[49]

This period was a tumultuous one for Nelson himself, not only because he had coldly abandoned Fanny and had suffered the loss of his dear friend Locker, who had recently died, but also because Emma gave birth to their illegitimate daughter, Horatia, on or around 29 January 1801. Still, at least there would be no further sinful infidelity, he reasoned, because he had now irrevocably separated from Fanny and was not, therefore, guilty of loving two women simultaneously. He loved only one: his Emma, who was "so true and loyal an Englishwoman," he rhapsodized to her, that she would naturally hate anyone who stood against "our King, Laws, Religion, and all that is dear to us."[50] It was this paragon of virtue with whom he now planned to share his life and to whom he swore "eternal fidelity."[51] Writing (in the third person) to Emma on 3 February 1801, he expressed hope that "the time may not be far distant when he may be united for ever to the object of his wishes, his only, *only* love. He swears before heaven that he will marry you as soon as it is possible, which he fervently prays may be soon."[52]

Emma was herself still married, however, which surely must have placed the issue of sin back in the forefront of Nelson's mind. Yet it does not appear to have concerned him as much as had his own unfaithfulness to Fanny. Apparently, Nelson understood the marriage between his mistress and her elderly husband to be a sexless union of friendship, based on filial affection (essentially like the love of a daughter and father) and not on romantic love and sexual intimacy. Thus, neither Nelson nor Emma was sexually involved with anyone else. They had only each other.

As Sir William was old and unwell, the admiral suspected that he might soon die, at which point, he told Emma, he "would instantly come and marry" her.[53] That is not to say he wished for Sir William's death. He cared for him greatly. In August 1801 he purchased Merton Place, a house for him to share with Emma and her ailing husband. As it happened, Sir William died, with Nelson and Emma holding his hands, in April 1803. With Sir William dead and with Fanny "out of sight and out of mind," Nelson and Emma were finally able to live as they had long wanted: together and with no one else, much to several of his closest friends' concern.[54]

After Nelson had irrevocably severed his ties with Fanny, yet before Sir William had passed away, he began privately addressing Emma as his "dearest wife."[55] In the sight of man, he apparently believed, their union was illegitimate; but in the sight of God, it was now more acceptable. In March 1801 he addressed a letter to Emma as "my own dear wife," adding "for such you are in my eyes and *in the face of heaven*."[56] He added a wish that the heavens would bless "my love, my darling angel, my heaven-given wife, the dearest only true wife." Clearly he had forgotten that he likewise once thanked God for uniting him and Fanny in matrimony.

He nonetheless remained tormented, and repeatedly asked God for peace, comfort, and a way of making into a righteous situation the relationship that even Emma privately admitted was still sinful: "I must sin on and love him more than ever; it is a crime worth going to Hell for."[57] "My God! My God!" he implored in a letter to Emma in March 1801, "look down and bless us; we will pray to thee for help and comfort, and to make our situation more happy."[58] He beseeched God, on a different occasion, to help and support him, even though he considered himself one of the Almighty's most "unworthy" servants.[59] All he wanted, he said in another letter to Emma, was to lead a quiet, sinless life with her and their daughter in the countryside: "Aye, how different I feel! A cottage, a plain joint of meat, and happiness, doing good to the poor, and setting an example of virtue and godliness."[60]

His awareness that he wasn't quite meeting God's expectations in this one area of his life only increased his desire to earn penance, if you like, by repeatedly throwing himself fully into the tasks he believed God had given him: protection of Britain and its king, and defense of the Christian religion and its established and proper societal norms. In this sense, at least, his adultery with Emma was of great national benefit.

It is worth noting Nelson's desire to "do good to the poor," mentioned above. He sincerely believed that caring for the poor was a central Christian duty, and, after receiving the Order of the Bath in 1797, he chose as the motto on his coat of arms the three words that summed up his "Christian walk" as well as his career: "Faith and Works" *(Fides et Opera)*.[61] Throughout his years at sea, he annually sent money or thoughtfully chosen gifts for the poor of Burnham, his childhood community.[62] One year these gifts included blankets "of the very best quality" to keep out the cold Norfolk winters.[63] "Good works" meant more than merely the giving of alms, however, and Nelson always demonstrated a desire to help those less fortunate, be it through paying the medical or funeral expenses of poor seamen or personally paying for the tuition of several hard-up midshipmen and junior officers. During his triumphant tour across England in the summer of 1802, he even left money at Oxford jail for the prisoners and a few days later, talked with and offered comfort to inmates in Gloucester prison.[64]

Nelson's good works perhaps satisfied his sense of obedience to one of God's central commands, but they and all his other acts of faith and Christian devotion still could not provide him with the inner peace he never quite achieved. He remained deliriously in love with Emma but remorseful at his sinful nature. How could it be otherwise? He wanted to please God and struggled hard to do so, but knew that his love for Emma was all-consuming. "My soul is God's," he wrote in October 1803; "let him dispose of it as it seemeth fit to his infinite wisdom, [but] my body is Emma's."[65]

In God's Hands

Nelson's statement that his soul belonged to God, who could take it away as and when He saw fit, represents one of the most remarkable and ultimately poignant aspects of Nelson's Christian faith. He sincerely believed that all issues of life and death belonged to God—a wondrous view for a warrior famous for his fiery aggression toward his enemies and his constant willingness to expose himself to life-threatening dangers. This belief grew not only from his birth as son of a clergyman and his apprenticeship under Locker and others, but also from an epiphany he experienced as a dreadfully ill seventeen-year-old in mid-1776. Sick with malaria, the young officer had what he frequently described thereafter as a visitation of a "radiant orb." It infused him with patriotism toward his king and country and with total confidence that he would survive and go on to play a role in God's great plan. "Well then," he later remembered thinking, "I will be a hero, and, confiding in Providence, I will brave every danger."[66]

The key words here are "confiding in Providence," *not,* as many writers suggest (focusing only on Nelson's remarkable courage and desire for fame), the words "I will be a hero." The verb *confide* now usually means to tell someone something of a private nature, but Nelson used it with a very different meaning in mind. He used it to denote having strong confidence or trust in someone or something. For example, shortly before the great clash at Trafalgar, Nelson proposed sending this signal to his fleet: "England confides that every man will do his duty."[67] This was a tricky message to send, given the system of flag communication then in use, so the word "confides" was changed to "expects," which was not quite the meaning Nelson wanted to convey.

In this way, then, Nelson was clearly linking his successes in battle and his boldness in action to his total confidence in the Almighty's guidance and divine intervention. From that moment of spiritual revelation during his illness in 1776 onward, he refused to take counsel of his fears, instead asking and trusting God to protect him. "I have great reason, my dearest Fanny, to be thankful to that Being, who has ever protected me in a most wonderful manner," he wrote home in April 1794.[68] Four years later he wrote to Sir William Hamilton before the anticipated battle against Napoleon's Egyptian invasion fleet: "God is good, and to Him do I commit myself and our Cause."[69] At about this time he told the same friend in another letter that he was determined to attack the enemy fleet regardless of how strong its defenses were, "and, with the blessing of Almighty God, I hope for a most glorious victory."[70]

This sense of the ever-watchful gaze of the Almighty, coupled with his desire for fame, filled Nelson not only with a strong sense of purpose, but also with tremendous physical and moral courage. In turn, that courage energized and inspired his men, who knew their captain, and later admiral, would not demand them to do or risk anything he had not done, or would not do, himself.

His strong belief in Providence and his sense that he had a role in God's plans does not mean that Nelson considered himself, or his captains and crews, immune from troubles or safe from danger. He reportedly prayed every day throughout his life and several times each day during times of crisis, asking for protection and a submissive heart should the Lord choose to let injury or misfortune touch them. He personally received many serious wounds, including the loss of vision in his right eye and the amputation of his right arm. Yet he accepted these wounds with remarkable resignation and humility. They happened because God wanted them to. After the blinding of Nelson's eye, his father wrote to reassure him that "it was an unerring power, wise and good, which diminished the force of the blow by which your eye was lost. . . . Your lot is cast, but the whole disposing thereof is of the Lord; the very hairs of your head are numbered—a most comfortable doctrine."[71] His son clearly agreed. Nelson gave public thanks to God for sparing his life and was often able in later years to gain humor or make a dramatic point because of his loss of sight. During the Battle of Copenhagen, for example, when he chose not to obey a signal sent by his overcautious commander in chief, Nelson reportedly placed his telescope to his blind eye and exclaimed: "I really do not see the signal."[72]

He responded to the amputation of his arm and his subsequent agony with the same Christian acceptance and humility. When he had recovered from the worst of his pain, he offered public thanks at a local church: "An Officer desires to return thanks to Almighty God for his perfect recovery from a severe wound, and also for many mercies bestowed him."[73] Rather than wallow in self-pity, he subsequently turned his amputation to his advantage, using it as a source of humor and even as a means of demonstrating to similarly disfigured seamen that he was, in essence, "one of them."

Characteristically, his first act after returning from Copenhagen in 1801 was to visit naval hospitals in Yarmouth. He came across one seaman who had suffered an almost identical amputation. Looking down at his own empty sleeve, he playfully said: "Well, Jack, then you and I are spoiled for [work as] fishermen; cheer up, my brave fellow."[74] This lighthearted act of empathy had a "magical effect" on the poor seaman, whose "eyes sparkled" as Nelson turned away. Word of this affectionate gesture spread fast, strengthening the already high esteem Nelson's men held him in, and it soon became part of the Nelson legend.

Nelson treated the prospect of death—not only that of his subordinates, but also his own—with the same submissive and humble spirit. Generals and admirals planned and conducted battles, but God, and God alone, decided who would live and who would not. The great seaman therefore repeatedly and fervently prayed for his men and himself, asking God to find them favorable. This request for divine protection and willingness to submit to God's will robbed death of its sting. In 1793 he jotted in his journal: "Though I know neither the time nor the manner of my death, I am not at all solicitous about it because I am sure that He knows them both, and that He will not fail to support and comfort me."[75] Similarly, during the Copenhagen campaign of 1801, he told Emma: "I own myself a BELIEVER IN GOD, and if I have any merit in not fearing death, it is because I feel that His power can shelter me when He pleases, and that I must fall whenever it is His good pleasure."[76] His most famous prayer of this kind, although he uttered and wrote down many, is found in his private diary for 13 September 1805:

> May the Great God whom I adore enable me to fulfil the expectations of my Country; and if it is His good pleasure that I should return, my thanks will never cease being offered up to the Throne of His Mercy. If it is His good providence to cut short my days upon earth, I bow with the greatest submission, relying that He will protect those so dear to me, that I may leave behind. His will be done: Amen, Amen, Amen.[77]

He met death five weeks later with the submission and humility for which he had prayed. On 21 October 1805, he realized battle was imminent and drafted another battle prayer (his last), in which he again committed his life to God and asked for "a great and glorious victory" that would involve no "misconduct" and end with British humanity to the vanquished. This was a prayer written for posterity (he made two copies to ensure that it would not be lost), but it was also genuinely meant as a plea to God. When his signal lieutenant, John Pasco, entered the admiral's cabin, he found Nelson still on his knees, reverently writing and speaking the prayer. Pasco remained standing "stationary and quiet until he arose . . . not wishing to distract his devotions."[78] Thus, Britain's greatest hero prayed on his knees in his cabin before ordering the commencement of his greatest battle. It bore fruit; his innovative plan worked marvelously, with the combined Franco-Spanish fleet suffering a humiliating and total defeat.

Killed by a sharpshooter at the height of the battle, Nelson died the way he always wanted: at the moment of victory in a great battle in the service of his country.[79] He lived long enough to learn that he had justified his nation's trust and permanently removed the threat of an invasion by Napoleon's forces. "Thank God I have done my duty," he sighed as he lay dying. He asked those present to ensure that Lady Hamilton and their daughter would be taken care of, and he prayed with his chaplain for divine mercy. Then, with his pain increasing and death approaching, his mind focused on the creator he soon anticipated meeting and to whom he had committed his soul. To Dr. Scott he whispered, almost as a rhetorical question: "I have not been a great sinner, doctor."[80] The last words caught by Scott, bending close to the admiral's lips, were the fitting: "God and my Country."[81] Despite his human imperfections, Nelson had served them both extremely well.

Nelson as Pastor

It would be misleading to suggest, despite his remarkable prayer life and its emotional and spiritual rewards, that Nelson's religiosity was only a personal matter between him and his creator, and that its primary external manifestation was the great courage and sense of purpose it gave him. These things were very important, and certainly proved inspiring to his captains and crews, who loved serving under him and swore they would follow him anywhere and meet whatever expectation he placed upon them. Yet Nelson's religious beliefs also had a significant impact on the way he commanded men, understood his pastoral duties as an officer, served his nation, and saw his allies and enemies.

Captain Locker provided Nelson with a splendid model of leadership, one that Nelson adopted and always retained. In particular, he realized that his "young gentlemen"—captains' servants and teenage midshipmen—most of whom had joined ships' companies at the age of twelve or thirteen, were still children who needed paternal guidance, counsel, and role models. Regardless of the systems then in place to train the junior officers in basic seamanship, their social, moral, and spiritual well-being was ultimately the responsibility of admirals and captains. Nelson, son of a strict but devout and loving father, and protégé of several devoted and benevolent officers, including his uncle and Captain Locker, always took this paternal responsibility seriously. He wanted his own "young gentlemen" to be good Christian warriors who would skillfully and zealously fight for God, king, and country, all the while being models of Christian virtue and honor to those who were, in turn, rising through the ranks beneath them as well as to the crews they would command.

Even as a young captain, Nelson proved unusually attentive to his midshipmen, calling them his "children" and devoting much time to their care and training. "All my children are well," he would write to friends and family members from the Leeward Islands in the mid-1780s,[82] even though he was himself considered by Prince William Henry (later crowned as King William IV) to be "the merest boy of a captain I ever beheld."[83] He took time out when he could to watch them take their school and nautical lessons, and believing in learning by example, he sometimes personally demonstrated how particular tasks were to be performed. On one occasion (before he lost an arm, of course) he even raced up the ratlines (the rope ladders) to the masthead, to demonstrate to a frightened youngster that he had nothing to fear.[84]

He also believed that his boys and midshipmen needed to be taught moral conduct, good social skills, and the appropriateness (or not) of certain types of behavior. As captain, he knew, he was the best role model. Early in his career, he developed the habit, which he retained even during later hectic campaigns, of always inviting a midshipman or two to breakfast. At Copenhagen in 1801, he apparently never invited fewer than four, as one youngster recalled with obvious appreciation.[85] Nelson delighted in their company and would even "enter into their boyish jokes."[86] They also needed outside, nonnaval role models. Lady Hughes remembered how, in 1784, Nelson took young midshipmen and lieutenants to official functions. "Your excellency must excuse me for bringing one of my midshipmen," she remembered Nelson saying on one occasion, "as I make it a rule to introduce them to all the good company I can, as they have few to look up to besides myself during the time they are at sea."[87]

These acts of kindness and good sense left lasting impressions on the young fellows, who remained ever after in awe of their captain. One awestruck young midshipman was William Hoste, who, soon after joining Nelson's ship at age twelve, proudly told his family that the paternal captain was "exceedingly kind to me."[88] In another letter the boy boasted that Nelson was "acknowledged to be one of the first [that is, best] captains in the service and is universally beloved by his men and officers."[89] Hoste became one of the many midshipmen that Nelson virtually adopted as his own children. In 1797 the admiral wrote to Hoste's father, telling him that "dear William" was very courageous and good and that "each day rivets him stronger to my heart."[90] Young Hoste repaid Nelson's affection. He proved a fine protégé and had a splendid naval career, distinguishing himself in the Battle of Lissa in 1811. Turning to his signal officer as the opposing squadrons closed for that great battle, Hoste ordered him to send an immediate signal to the squadron. It read: "REMEMBER NELSON."

Nelson did not neglect the spiritual training of his young wards, or of his entire crews for that matter. A regular churchgoer himself during periods of shore leave (as were many of his famous "band of brothers," including Sir James Saumarez, Ball, and Parker), he insisted that, whenever weather permitted, all available crews assemble on deck for divine services on Sundays.[91] Even when pursuing the French fleet across the Atlantic and back before the climactic Trafalgar campaign in 1805, Nelson had his crews attend services. Moreover, he made a point of discussing the sermons afterward with the chaplain, thanking him and humbly offering suggestions for future sermons or other ways of conveying certain messages. The admiral was "always anxious," his chaplain's biographer wrote, "that the discourse should be sufficiently plain for the men."[92] Nelson sometimes even took down from his shelves "a volume of sermons in his own cabin, with the page already marked at some discourse, which he thought well suited to such a congregation, and requested Dr. Scott to preach it on the following Sunday."

Never embarrassed to acknowledge what he perceived as divine assistance, Nelson was often seen to thank God for positive circumstances. Before the Battle of Copenhagen, in open view, "he fervently thanked his God" for sending a favorable wind.[93] After individual engagements and even squadron and fleet battles, he always had his crews assembled on deck to offer public thanks to God. After the stunning victory over the French in the Battle of the Nile, he sent this memorandum to his captains: "Almighty God having blessed His Majesty's Arms with Victory, the Admiral intends returning Public Thanksgiving for the same at two o'clock this day; and he recommends every ship doing the same as soon as convenient."[94] One seaman later described such after-battle thanksgiving services as the "Nelsonian style."[95] That is not to say that all seamen understood or shared the sentiments. George Charles Smith, who later founded the London Mariners' Church, recalled that relatively few of the hard-living seamen were practicing Christians. Before the Battle of Copenhagen, the nineteen-year-old Smith climbed secretly to the foretop to pray, "afraid and ashamed to be seen on my knees, as every man would have ridiculed me, and called out *Methodist,* which at that time was a most contemptable [*sic*] word, and meant coward."[96]

Nelson's commitment to the spiritual well-being of his crews did not end with divine services and prayers of thanksgiving. He also wanted those who could read to study and understand scriptures (illiteracy was not uncommon among seamen in the navy).[97] Before leaving for sea in January 1801, he repeated what he had done before several previous voyages.[98] He wrote to the Society for Promoting Christian Knowledge, founded 101 years earlier, and gently reminded this organization, which granted books to naval forces and armies as well as to evangelical clergymen in impoverished areas, that he had previously obtained their bibles for the crews of the *Agamemnon* and the *Vanguard*.[99] He added (perhaps optimistically, given Smith's testimony that religion was not a great force in the navy) that the crews' subsequent conduct was hopefully such as "to induce a belief that good to our King and Country may have risen from the Seamen and Marines having been taught to respect the Established Religion."[100] He thus humbly asked that the society would present a further consignment of bibles and prayer books to the crew ("the no. near 900") of the *San José* and expressed a hope that the ship would be as "successful as the former ships you gave them to."

"Success," of course, meant righteous conduct, Christian worship (and hopefully belief), and victory in battle. These would, Nelson trusted, bring him and his crews honor and hopefully some prize money, but also provide necessary tangible evidence of their Christian responsibility to king and community. His thankful prayers that God had given victory to "His Majesty's Arms"—and he always described his victories as ultimately the king's—are testament to his unswerving loyalty to his king and country.

Society, he believed, rested upon the necessity of subordination and the just demands of authority. God had chosen monarchy as His means of imposing authority on society; and, therefore, devotion to the king was the moral obligation of faithful Christians. This line of reasoning naturally led the admiral to despise republicanism, especially the French Jacobin variant with its antimonarchical and atheistic principles. These principles, Nelson wrote to the duke of Clarence (referring to republican clubs springing up in England), are based "on principles certainly inimical to our present constitution both in church and state."[101] His attachment to the king, Nelson stressed, would naturally never be shaken, and would only end with his death.

Nelson thus reasoned that the enemies of this divinely inspired order—and for most of his life this meant the French—were also the enemies of God and the king. They must be opposed with all possible effort. Nelson burned with passion for God and the king, and with hatred of their enemies, and he wanted his young charges to feel likewise. He told one new midshipman whose father reportedly had Whiggish and possibly even revolutionary views: "There are three things, young gentleman, which you are constantly to bear in mind: first, you must always implicitly obey orders . . . secondly, you must consider every man as your enemy who speaks ill of your King; and thirdly, you must hate a Frenchman as you do the devil."[102]

Nelson naturally served his own king, George III, with the loyalty of a devoted subject, but he also treated foreign monarchs with great respect and hoped that, united against those who wanted to dethrone them, they would "send all Republicans to the devil."[103] In Nelson's mind, foreign monarchs—even Catholic kings, such as Louis XVI of France (who had been guillotined in 1793) and Ferdinand IV of the Kingdom of the Two Sicilies—were divinely appointed governors of their peoples and therefore worthy allies in the war against the "irreligious" French. After receiving a title and a Sicilian estate from Ferdinand, a "lawful sovereign," in return for helping him regain his kingdom from Jacobin revolutionaries, Nelson wrote to thank him. He signed himself "your Majesty's faithful servant."[104] Expressing belief in Ferdinand's divine appointment he added a prayer that went further than the pro forma courtesies required, hoping that "the Almighty may pour down his choicest blessings on your sacred person, and on those of the Queen and the whole Royal Family."

The English hero was thoroughly devoted to the established church of his nation and was a Protestant by both inclination and practice. Yet he showed no discomfort with other expressions of faith, such as the Islamic religion of his Turkish allies and others. As will be shown below, Nelson appreciated any and all spiritual devotion and felt infinitely more offended by godless peoples than he did by either Muslims or Catholics.

He appreciated the Neapolitans' Catholicism and certainly never let religious differences interfere with his zealous diplomatic efforts to keep Naples on Britain's side in the war. He was always respectful to Catholic dignitaries and went to great lengths to ensure that they knew he would defend their Church and its role in Catholic nations.[105] He also demonstrated tolerance in a number of small-scale ways. For instance, he willingly acceded to a request from the queen of Naples for the discharge of one of Capt. Thomas Masterman Hardy's marines, who was a Catholic priest by training, so that he could serve in her kingdom.[106] And when in 1804 he gave gifts to the Catholic residents of the Maddalena Islands to repay their hospitality, he chose church plate (a silver crucifix and candlesticks) for the local parish, a gesture that prompted the startled parish priest to cut short a trip elsewhere so that he could thank Nelson personally. The priest even promised, in a gushy letter, ever after to offer daily vows for Nelson's long life, prosperity, and glory.[107]

Nelson also bemoaned that the frail and unwell Pope Pius VI, whom he had often hoped would join the war against atheistic France, had been deported to Valence (where he later died). It "makes my heart bleed," Nelson lamented.[108] When the new pope, Pius VII, returned to Rome in 1800, Nelson wrote him a congratulatory letter. He explained that he had himself played a part in making the Catholic Italian states safe for the pope's return.[109] "Holy Father," he added, "I presume to offer my most sincere congratulations on this occasion; and with my most fervent wishes and prayers that your residence may be blessed with health, and every comfort this world can afford." This was no mere polite letter to a distinguished personage to praise him for good fortune. The letter had a curious religious purpose; the admiral felt compelled to tell the pope that in 1798 a priest had predicted Nelson would, by providing naval assistance, play a key role in Rome's recapture from the French. This had "turned out so exactly," Nelson said, that he felt the pope should know about the "extraordinary" consonance between the prediction and the outcome.

It is hard to say whether the admiral saw this as the fulfillment of a prophecy—at the time he said he didn't[110]—or whether he was merely trying to convey to someone who would understand that the Lord had indeed selected Nelson to do something truly worthwhile and perhaps a little miraculous. Horatio Nelson often gained amusement or wonder from God's way of operating. The best example, and a fine one with which to end this chapter, is his description of his tactical briefing with his captains before the Battle of Trafalgar. Although he would never blasphemously compare himself to his savior, he was struck by how much his meal and briefing with his confidants—who loved him as disciples—resembled Christ's last supper. "When I came to explain to them the 'Nelson touch,'" he wrote to Emma, "it was like an electric shock. Some shed tears, all approved."[111] His confidants, perhaps realizing the perilous nature of the tactics and sensing their leader's belief that he would not survive, reassured him of their devotion. Nelson was much affected. "Some may be Judas's," he told Emma, continuing with the Last Supper symbolism, "but the majority are much pleased with my commanding them."

They were indeed, and his death shortly thereafter initiated the cult of Nelson that has survived the last two hundred years. It has not raised the admiral to the status of deity, of course, but it has given him an immortality of sorts. His reputation as England's greatest hero is secure and, far from diminishing with time, seems only to grow.

Conclusions

Nelson had devoted virtually his entire life, from twelve-year-old midshipman to forty-seven-year-old vice admiral, to his great loves: God, his country, and the navy. For them he sacrificed blood, an eye, an arm, and finally his life. He was devout and passionately, openly Christian. Yet he was no saint. His vanity was legendary, as was his craving for fame. The main factor mitigating against his egotism, which often prompted foolish behavior, was his total lack of malevolence. Unlike Napoleon Bonaparte, the other larger-than-life figure from the period, Nelson never sought to dominate, exploit, or overpower people. Aside from his behavior during battle, which transformed him into a firebrand, or when he was grumpy because of seasickness or fatigue, Nelson was usually gentle, encouraging, and generous. He was quite capable of weeping when friends and family—or even seamen on his ships—suffered misfortune. His six-year affair with Lady Hamilton, wife of a British diplomat, also damages his otherwise spotless reputation, especially as the great love between them soon led him to separate permanently, and in unpleasant circumstances, from his long-suffering wife, Fanny. Yet even that departure from the Christian values he extolled is balanced by the fact that he remained deeply, rapturously in love with Emma, and she with him. In God's eyes, he hoped, their union was not dreadfully wicked. "I have not been a *great* sinner," he told himself shortly after repenting of his sins one last time before the sleep of death closed his eyes forever.

As it is, Nelson's flaws and poor choices only show him to be typically human, capable of making mistakes like anyone else. The British people realized this, and delighted in the fact that their greatest hero was one of them. Nelson also recognized his human imperfections and weaknesses. He made peace with himself and God, as well as one can, by praying for protection and forgiveness. He also tried hard to meet God's expectations and stayed ever thankful for the talents as a warrior and the opportunities to use them that God had bestowed upon him. For over two hundred years now, Britain has felt the same gratitude. And for good reason: no other Briton, with the closest competitor being Sir Winston Churchill, has done more to inspire the nation to unite against tyranny. The dukes of Marlborough and Wellington certainly won splendid triumphs, for which they were richly rewarded; but they were aristocrats with whom ordinary English men and women could not identify. And they never raised national morale and inspired the British people to fight for victory to the degree that Nelson did.

Vice Admiral Lord Viscount Nelson, Britain's paradigmatic hero.

Commencement of the Battle of the Nile, 1 August 1798.

The Battle of Copenhagen, 2 April 1801.

Nelson accepts the swords of the vanquished, 14 February 1797.
© National Maritime Museum, Greenwich. Photo PAD 5540

Leading from the front in order to inflame passion and inspire loyalty, the adrenaline-charged Nelson attacks a Spanish launch, July 1797.

The horrors of sea battle at night, 1 August 1798. *L'Orient* explodes with hideous ferocity during the Battle of the Nile.

The cataclysmic Battle of Trafalgar, 21 October 1805.

Naval crews pulled their ships' guns and ammunition ashore on Corsica, then dragged the heavy ordnance miles overland to the batteries they created and kept supplied by making continuous return trips to the ships.

3

Command, Leadership, and Management

LORD NELSON has a grand and richly deserved reputation as an inspiring and inspired leader of men, who earned the unswerving devotion of his captains and crews and molded them into the most ferocious and effective naval force of the eighteenth and nineteenth centuries. His leadership talents have been lauded in popular and academic articles and books, in specialist military encyclopedias and reference works, and even in several television series. One example of the latter is the *Great Commanders,* a popular British Channel 4 series narrating the careers of history's six best commanders. It includes Nelson as its only seaman, the others—Alexander, Caesar, Napoleon, Grant, and Zhukov—being soldiers.[1] Nelson's inclusion in this esteemed pantheon stemmed not only from the victories he gained, which rival in brilliance and significance those of his legendary fellows, but also from his seemingly unparalleled ability to inspire loyalty, commitment, and courage.

Yet, the ability to inspire and motivate is not the sum total of warfighting *command,* an umbrella term now understood to encompass three separate but interrelated functions:[2]

1. *command,* which denotes (a) the legally vested rights of an individual within a structured hierarchy to exercise authority over subordinates by virtue of rank or appointment; (b) the maintenance of that hierarchy

or command structure through discipline; and (c) the assumption of responsibility for subordinates' safety, welfare, and morale;

2. *leadership,* which denotes the art of influencing and persuading those assigned to particular goals to work willingly, energetically, and in unison toward the attainment of those goals;
3. *management,* which denotes the process of establishing task priorities and of allocating and directing time and resources efficiently so that subordinates can carry out those tasks effectively.

This conceptual separation of functions is relatively new, and one cannot of course suggest that Nelson, or any other commander of his era, consciously understood his responsibilities in such a conceptually sophisticated fashion. Yet even a cursory study of his career reveals that Nelson exercised, often instinctively and without much deliberation, virtually all the functions that are now considered integral to command, to leadership, and to management.

Nelson's leadership has been analyzed frequently and adequately. Still, the way he understood and exercised his authority in the other two spheres —command and management—have been relatively ignored. This chapter seeks to redress the imbalance by analyzing how the great admiral saw and performed his roles as commander, leader, and manager. It also asks whether his instinctive and unparalleled talent for inspiring and motivating was matched by an ability to exercise authority judiciously and manage time, men, and matériel efficiently and effectively.

Command: Nelson and His Superiors

Using the definition of command given above—that it is a person's legally vested right to exercise authority over subordinates by virtue of rank or appointment, and that person's responsibility to ensure subordinates' safety, welfare, and morale—it becomes clear that Lord Nelson's exercise of command functions has been inadequately explained. It has been overshadowed by what writers have clearly found more exciting and dramatic: his charismatic leadership style and his highly developed operational and tactical abilities. Nevertheless, from the tender age of twenty, when he received the rank of commander and then of captain, Nelson successively assumed sole authority over his own crews, squadrons, and fleets as well as the daunting responsibility to keep them safe, healthy, happy, and capable of effective warfighting.

Nelson always respected the Royal Navy's system of hierarchical control, which was administered by the Admiralty but exercised by operational, sea-going commanders with minimum interference from London. This system placed greater authority and responsibility in the hands of each successively senior seaman and officer, and gave admirals and captains at sea considerable independence in their use of authority. Put simply, it was the duty of every seaman and officer to obey his superiors' instructions without question or delay, and to comply fully with the thirty-six Articles of War that provided the legal basis for naval discipline.[3] These Articles, each between 50 and 150 words long, did not define the duties and roles of seamen and officers, but laid down a wide range of binding rules and regulations for correct conduct. They also specified punishments for unacceptable conduct (including the death penalty for serious offenses). In this way, the Articles served as a practical but unsophisticated law code for sea service.

Fortunate to have served under a series of capable and committed captains who ran their ships with fairness and a close eye on discipline and morale, the young Nelson learned early the importance of submission to authority and respect for his superiors. His delighted mentors recognized this characteristic and trusted him with increasingly important responsibilities. He embraced these eagerly and his devotion to duty, righteous conduct, and success with tasks earned him quick promotion. At the time of his appointment to the thirty-two-gun frigate *Lowestoffe* on 10 April 1777, a day after the eighteen-year-old's confirmation as lieutenant, his uncle, Capt. Maurice Suckling, gave him a set of instructions to help guide him throughout his career. It opened with advice that Nelson would always try to follow: "My dear Horatio, Pay every respect to your superior officers, [in the same way] as you shall wish to receive respect yourself."[4]

Nelson ceaselessly demonstrated courteous respect for his superiors and never treated any, even those he criticized in his private letters, with visible disdain. He naturally wanted to be respected himself and always worked hard to act energetically, properly, sensitively, and generously. It paid off. One of his peers noticed in 1796 how well Nelson interacted with his superiors, and he jealously claimed that this resulted in Nelson's receiving better assignments than other captains. "You did just as you pleased in Lord Hood's time," the captain complained, "the same in Admiral Hotham's and now again with Sir John Jervis; it makes no difference to you who is commander-in-chief."[5] Nelson gave this petty-minded captain short shrift.

There were exceptions to his cordial relationships with superiors, however. Nelson could never form any bond of respect or friendship with commanding officers who were not as devoted to their nation's interests as he thought they should be. He felt deeply annoyed, for example, to learn that his zealous enforcement of the Navigation Acts during his service in the Leeward Islands between 1784 and 1787 had earned him the hostility of local islanders. Even worse, it earned him criticism from Rear Adm. Sir Richard Hughes, his squadron commander, as well as from other naval superiors.

He indignantly argued that by suppressing illegal trade between American merchantmen and British colonists, he was only obeying the letter of the law. This point was technically correct but ignored that Admiral Hughes had encouraged the young captain to obey the *spirit* of the law without being over zealous in his devotion to what the young captain considered his "solemn duty." In March 1785 Nelson explained the reason for his apparent stubbornness to Lord Sydney, the secretary of state: "My character as a man," Nelson said, "will bear the strictest investigation. . . . I stand for myself; no great [family or political] connexion to support me if inclined to fall; therefore my good name as a Man, an Officer and an Englishman, I must be very careful of. My greatest pride is to discharge my duty faithfully; my greatest ambition to receive approbation for my conduct."[6]

In truth, Nelson was not being disobedient to his superiors, just excessively legalistic, and the frustrated Admiralty eventually accepted that he was not acting improperly. He nonetheless quarreled with Admiral Hughes over another matter during this period: Nelson's refusal to take orders from Capt. John Moutray, the commissioner at Antigua, who, although retired from active service on half pay, still considered himself Nelson's superior. Nelson, again technically correct but unnecessarily inflexible and legalistic, thought otherwise. He responded to Hughes's anger by maintaining that he would obey Moutray if Moutray were ever reappointed to active service, and would certainly "not entertain a doubt upon the subject."[7] Not one of his superiors, he added in his February 1785 letter, should "have the smallest idea that I shall ever dispute the orders of my superiors." He would never do so. He made this same assertion to his future wife, Fanny, a year later in March 1786, when discussing another matter. "I can't bear the idea of disobeying orders," he wrote; "I should not like to have mine disobeyed."[8]

This comment may bring smiles to historians (blessed with the luxury of hindsight), who know—as the British public soon would—that Nelson would famously disobey superior orders on three occasions. The first two times earned him extra popularity with the public and the Admiralty, and the last earned him criticism from his superiors for his poor judgment.

The first two occasions, of course, were the Battle of Cape St. Vincent on 14 February 1797 and the Battle of Copenhagen on 2 April 1801. In the first battle, Nelson spontaneously wore his ship out of the British line and attacked the Spanish van as it attempted to escape. He did so on his own initiative, knowing that he had not received instructions or permission from his commander, Sir John Jervis, and that his decision to leave the British line violated tactical principles laid down over a century earlier and accepted virtually as gospel ever since. His bold move secured victory and Jervis's profound thanks. In the second case, at Copenhagen, he directly disobeyed Sir Hyde Parker's order to break off battle with the Danes because Nelson knew that if he just held on a little longer, he would win the war of nerves. He was right; victory quickly followed.

We have also seen that in July 1799 Nelson disobeyed his immediate superior, Admiral Lord Keith, who ordered him to remove his vessels from Naples, the Levant, and Malta and concentrate them around Minorca. Nelson, deciding that Naples and Sicily still needed his protection (a priority established in earlier general instructions from St. Vincent),[9] and that the ten-month blockade of Malta could not be abandoned, initially chose to ignore Keith's order. He only partially carried it out when Keith promptly repeated it. This action earned him no glory but the criticism of the Admiralty.

One can make a strong case for Nelson's apparent disobedience in the Battles of Cape St. Vincent and Copenhagen. His daring acts of initiative resulted from his awareness that through the confusing "fog of war," he could see better than his superiors how the battles were unfolding and what actions were needed. Although technically disobedient, he was in fact still acting in accord with his superiors' overall battle objectives.

One cannot so easily defend his defiance of Lord Keith in 1799. Nelson was entranced by his illustrious company at the Court of Naples, unwilling to leave Emma Hamilton, puffed up with pride at his recent victory at the Nile, and busy with several critical naval operations simultaneously. He therefore chose to establish his own operational priorities rather than submit, as he should have, to the priorities set by his lawful superior officer.[10]

Nelson doubtless honestly and sincerely believed he could not leave Naples in jeopardy at what he considered a crucial time, and felt equally unable to abandon prematurely and unsuccessfully his ten-month blockade of Malta, which seemed to be nearing an end. Yet these sincere beliefs do not represent an adequate reason for disobeying formal and explicit orders. He should have obeyed Lord Keith's orders promptly, even if only partially at first. Nelson should have sent half his ships until he could disengage the others (he had done precisely this on an earlier occasion, in May).[11] Meanwhile, he should have written to brief Keith fully on the situation in and around Naples and the eastern Mediterranean. But there are mitigating factors: Nelson had the best interests of the king and queen of Naples at heart; he had already restored their kingdom; he believed that the blockade of Malta was almost at an end and that the Russians would take Malta if the British blockade ended prematurely; and he meant no challenge to naval authority or personal disrespect to Lord Keith, whom he generally found agreeable.

Indeed, one must concede that Nelson was a remarkably loyal and conscientious officer, who always had his nation's well-being and honor at the forefront of his mind. If he occasionally felt compelled to act contrary to orders or bucked the chain of command, it was because he sincerely believed he possessed a clearer view of the immediate situation than his superiors and therefore had a solemn duty to follow the best course of action. He once referred to this belief as "political courage," noting that an officer must always do what he considered "right and proper for our King and Country's service."[12] This political courage was "as highly necessary as battle courage."

Nelson certainly never acted contrary to orders or challenged the chain of command with casual disinterest. He always fretted over the way his actions would be interpreted. After his victory in the Battle of Copenhagen he said to those close to him: "Well! I have fought contrary to orders, and I shall perhaps be hanged."[13] Although he considered his victory to be ample justification for disobeying Parker's signal (it was!), he added: "never mind, let them."

Command: Nelson and His Subordinates

Nelson expected total and immediate obedience from his *own* subordinates, even though he seldom had to articulate this assumption because of his immense popularity among officers and seamen. Recognizing his extraordinary tactical skills and his compassion and care for them, they wanted to serve under him and readily and unquestioningly accepted his authority. He considered this level of commitment and loyalty essential in a subordinate, and (with delicious hypocrisy) told one new midshipman that the first rule to remember was that "you must always implicitly obey orders without attempting to form any opinion of your own respecting their propriety."[14] Soon after meeting Edward Berry, with whom he would quickly form a close and lasting friendship, he told Sir John Jervis: "I have, as far as I have seen, every reason to be satisfied with him both as a gentleman and an officer."[15] Although one of Berry's actions miscarried through no fault of his own, Nelson emphatically added, "I was much pleased by Mr. Berry's strict attention to my instructions."

During the great fleet mutinies of June 1797, Nelson's own commander in chief, the earl of St. Vincent, responded to seamen's petitions and complaints with quick mutiny trials and immediate hangings.[16] Executions were ordinarily quite rare and were carried out according to the strictest criteria, but the navy's loyalty crisis of 1797 was so severe that Jervis, a strict disciplinarian at the best of times, felt compelled to act with swift brutality. Nelson himself had few difficulties with the loyalty and obedience of his crew on the *Theseus*. He and his flag captain, Ralph Miller, one day actually found a remarkable, revealing note from the crew:

> Success attend Admiral Nelson! God bless Captain Miller! We thank them for the officers they have placed over us. We are happy and comfortable, and will shed every drop of blood in our veins to support them, and the name of the *Theseus* shall be immortalised as high as the *Captain*'s [Nelson's previous ship].—SHIP'S COMPANY.[17]

Nelson nonetheless expressed support for St. Vincent's executions elsewhere in the fleet, seeing them as "severe examples" that were "absolutely necessary."[18] When St. Vincent received criticism for executing several mutineers on a Sunday, the Christian sabbath, Nelson reassured him that he fully agreed with the measure. Had it been Christmas Day, Nelson told one shocked colleague, he himself would still have executed them.[19]

He probably would *not* have executed seamen on Christmas or on a Sunday; he was as devout a Christian as any who lodged complaints with St. Vincent. Yet his desire to show respect for his lawful commander's authority and to convince others that such respect for the chain of command was essential compelled him to support St. Vincent. When Nelson was away from the watchful eyes of superiors, and particularly when he commanded his *own* squadrons and fleets, his devout Christianity often moderated his strict enforcement of the Articles of War.

In 1787, for instance, he presided over a court martial that condemned a deserter to death by hanging (in accordance with the articles). Asked to grant a reprieve, Nelson acceded to the request and discharged the miscreant from naval service with a pardon. Censured by the Admiralty for exceeding his lawful authority, he defended himself: "Thus was I near, if not cutting the thread of life, at least of shortening a fellow creature's days. The law might not have supposed me guilty of murder, but my feelings would nearly have been the same."[20] The Admiralty was unimpressed, arguing that Nelson had no right to discharge the pardoned man, much less without punishment. This prompted Nelson to explain, using Christian imagery of sin, repentance, and grace: "I had also always understood that when a man was condemned to suffer death, he was from that moment dead in Law; and if he was pardoned, he became as a new man."[21] The man could, Nelson added, then choose whether to reenlist or not.

Similarly, in 1803 Nelson was gravely disappointed to learn that some of the seamen in his fleet found blockade duties off Toulon so tough and monotonous that they consequently deserted. He considered desertion a personal affront and a miserable dereliction of duty to God, king, and country. He nonetheless publicly forgave the first batch of deserters in the hope that his Christian mercy would demonstrate his concern for his crew's well-being. It was only when further desertions occurred that he felt compelled to threaten the most dire consequences. On 7 November 1803, he sent the following memorandum to the captains and commanders of his fleet:

> Lord Nelson is very sorry to find that, notwithstanding his forgiveness of the men who deserted in Spain, it has failed to have its proper effect, and there are still men who so far forget their duty to their King and Country, as to desert the Service at a time when every man in England is in arms to defend it against the French. Therefore Lord Nelson desires that it may be perfectly understood, that if any man be so infamous as to desert from the Service in future, he will not only be brought to a Court-Martial, but that if the sentence should be *Death,* it will be most assuredly carried into execution.[22]

Nelson ordinarily accomplished the type of discipline he wanted without recourse to threats of severe punishments or execution. In 1797 after one entire crew in his squadron refused to accept the authority of a new captain because of that officer's reputation for excessive harshness, they threatened to mutiny. Promises of hangings from other officers only increased their rage, prompting them to take up weapons. A standoff ensued. Commodore Nelson ended it by calmly going aboard their frigate, calling all rebellious crewmen together, and gently asking them why they had taken such a drastic course of action.[23] They explained that they felt fearful of the new captain and would not submit to his command authority. Nelson threatened no punishment, but merely told his "lads" that they were a marvelous crew, had engaged and defeated two enemy frigates of greater strength, and, as the best frigate crew in the navy, should not now be rebelling. If the new captain treated them unfairly, they had only to bring a complaint to him and he would support them.[24] The crew erupted in cheers, swore loyalty to the new captain (who shed tears), and never caused further trouble.

This willingness to try a gentler form of discipline before the full weight of the Articles of War became necessary does not mean that Nelson was lax in his command responsibilities and cared little for day-to-day discipline (despite St. Vincent's odd claim in 1800 that Nelson's ship was "always in the most dreadful disorder").[25] On the contrary, although Nelson disliked having men flogged for minor offenses, he never hesitated to have them punished, sometimes severely, for wrongdoing if his attempts to modify their behavior by other means failed.[26] He was especially harsh on anyone making mutinous statements. As an admiral, he left the day-to-day exercise of command over crews to his ships' captains. One or two, such as Captain Berry, seldom resorted to flogging, whereas the famous Captain Hardy, Nelson's most trusted captain, used the lash more frequently. Nelson was unconcerned, believing that respect for command authority was far more important than his flag captains' actual means of obtaining it.

On matters of command authority and discipline, Nelson saw no difference between seamen and officers. All had to respect the chain of command and none could escape the consequences of not doing so. He felt particularly disappointed when misbehaving officers attempted to escape punishment by having their captains try to persuade him to overlook the offense or at least be lenient. In one instance in 1804, Nelson told a subordinate admiral that he would indeed waive a court martial, but only if the erring young officer would write a public letter of contrition and promise sincerely never to repeat the offense.[27] This letter would be read aloud to all the officers in the fleet and placed in its discipline registers. Nelson nonetheless rebuked the admiral who had interceded on behalf of his officer. "We must recollect," he said, that "it was upon the quarter-deck, in the face of the ship's company, that he treated his captain with contempt; and I am in duty bound to support the authority and consequence of every officer under my command."[28] He reminded the admiral: "A poor ignorant Seaman is for ever punished for contempt to *his* superior."

This gesture doubtless delighted Nelson's seamen, who felt their commander cared for them as much as for the officers. And indeed he did. Nelson —like many other senior officers, of course—always took his command responsibilities seriously, recognizing that ultimately it was up to him to keep his crews happy, safe, and healthy. Conditions at sea were seldom conducive to high morale among the crews, as Nelson was fully aware, but he did his best to convince his crews that they were valuable to him as people, not merely as workers. This was not difficult; unlike many fellow captains and admirals, he genuinely treasured his officers and seamen. Whereas his compatriot the duke of Wellington clearly did not like his soldiers,[29] Nelson delighted in his seamen. After he received command of the *Albemarle* and obtained a crew in 1781, he excitedly told his brother: "I have an exceeding good Ship's company. Not a man or officer in her I would wish to change."[30] This was his typical response to each of his new crews. After he received command of the *Agamemnon* in 1793, for instance, he wrote repeatedly and enthusiastically to family and friends about his marvelous officers and seamen. In 1800 he gushed with adoration of his crew aboard the *Foudroyant*. Nelson told the first lord of the Admiralty, George John Spencer, second Earl Spencer, that he loved this excellent crew as a "fond father" loves "a darling child."[31]

Nelson's devotion to his "lads," as he often called his subordinates, was not contrived. It was sincere and profound. After his arm was shattered during the 1797 raid on Tenerife, he ordered the barge that was urgently taking him back to his ship (for the brutal amputation that would save his life) to stop and rescue survivors from the sinking cutter *Fox*.[32] Likewise, after the

Battle of the Nile in 1798, his officers tried to get him immediate medical attention for an ugly wound to his forehead. He protested, insisting: "No, I will take my turn with my brave fellows."[33] These acts became known almost immediately throughout his crews, who marveled at his comradeship.

He recognized early in his career that his "brave fellows" received little recompense for their exhausting labors and willingness to risk mortal peril. Aside from a sense of pride if they contributed to victorious engagements or battles, they received meager wages and even these were paid sporadically. Prize money, on the other hand, could give them more material pleasure and a little to take home to their long-suffering families.

Nelson, who enjoyed receiving prize money himself but never pursued it at the expense of opportunities for glory, always went to great lengths to ensure that his crews received, promptly and fully, all wages and prize moneys owing to them. As a young captain in 1783, having just had his ship *Albemarle* paid off, he spent three weeks in London pestering the Admiralty for the "wages due to my good fellows."[34] He could not understand, he told his friend Locker, why it was necessary to pay off crews after each period of service, sending them off to serve under other officers and captains on different ships. This "infernal" system, he claimed in surprisingly modern terms, prevented the healthy bonding between officers and crews that made mutual respect possible.[35]

Even after he had achieved significant fame, he retained his commitment to his crews' well-being and happiness. After the Battle of the Nile he publicly thanked his crews for "their very gallant behavior in this glorious battle."[36] Every British seaman must surely know, he said in his letter of thanks, how superior they were "in discipline and good order" to the riotous French. He concluded by promising that he would not fail "to represent their truly meritorious conduct in the strongest terms to the Commander-in-Chief." He kept his word, informing St. Vincent that he owed his victory to all the men in the squadron, who had distinguished themselves through their discipline, commitment, and remarkable courage.[37] He then made himself quite unpopular with Earl Spencer by requesting an additional £60,000 prize money (of which his own share would be a mere £625) for three unseaworthy French ships he had to set on fire after the battle.[38] Although he was personally content knowing the Admiralty was pleased with him for his victory, he asked

Earl Spencer: "What reward have the inferior Officers and men but the value of the Prizes?" If crews' rewards were not paid, he insisted, an admiral could not expect their support.

Their support, he realized, depended not only on their happiness, but also on their mental and physical health. To promote mental well-being during a long and exhausting period of blockade in 1803 and 1804, he had his ships alternate their cruising grounds so that the crews would not suffer dreadful monotony but have different coastlines to look at.[39] He also promoted "cheerfulness amongst the men . . . by music and dancing and theatrical amusements," instituting these forms of activity on his flagship and encouraging them throughout his fleet.[40] Not neglecting the spiritual dimension of their well-being, he also encouraged participation in Christian services and procured bibles and prayer books for those who could read.

Crews' physical welfare was also important to Nelson, who had in early years lamented his inability—often because of logistical constraints caused by rapidly changing tactical situations, and sometimes because of the relative unconcern of some superiors—to keep crews healthy. On 6 February 1794, for example, Nelson glumly reported to his immediate superior, Commo. Robert Linzee, that he would follow orders and land a troop of soldiers then being transported to Corsica, even though this activity would leave his ship, the *Agamemnon,* unable to remain operational. "If you are pleased to look at our sick list," he said, "you will see our wretched state, [and that we will be] absolutely unable to keep the sea if so great a part of our strength is taken from us."[41] The clear inference from this letter is that he had so many seamen sick that the soldiers had been forced to do sea duties.

When he gained his own squadrons and fleets, he made health a priority. "The great thing in all Military Service is health," he told one doctor friend in March 1804, "and you will agree with me, that it is easier for an Officer to keep men healthy, than for a Physician to cure them."[42] Nelson insisted on stoves and ventilators being used below decks to minimize the negative effects of the ever-present damp and mildew. The additional appointments would also keep crews' clothes dry and their living areas at least a little warmer than outside temperatures.

He was not, of course, the first to care for the physical health of his seamen and he doubtless learned much from his own mentors. St. Vincent, whom Nelson deeply respected, always placed great emphasis on crews' health and insisted on decent food—particularly fresh meat, limes, and onions—to prevent the scurvy and other diseases and ailments that debilitated fleets during the great age of sail. Indeed, when it came to food, St. Vincent went even further than Nelson would, forbidding officers to receive choice cuts of meat or any other preference in terms of food.[43] Nelson did, though, go to extraordinary lengths to ensure that his ships were well provisioned, as will be shown below in a discussion of his management and logistical skills.

Leadership

Sir Michael Howard, one of the world's finest military historians, recently summed up leadership as "the capacity to inspire and motivate; to persuade people willingly to endure hardships, usually prolonged, and incur dangers, usually acute, that if left to themselves they would do their utmost to avoid."[44] This superb definition provides a useful frame of reference with which to study Nelson's relationship with subordinates. It is clear that he was not an *institutional* leader who relied heavily on the legal authority that came with his rank and appointment, or an *authoritarian* leader who relied on his ability to intimidate.[45] Unfortunately, these two command types overpopulate military history, with the duke of Wellington, Horatio Kitchener, several senior British generals of World War I, John Jellicoe, Arthur Tedder, and Dwight Eisenhower typifying the first type; and Douglas Haig, Bernard Montgomery, Douglas MacArthur, and Norman Schwarzkopf typifying the second. That is not to say that senior commanders with the aforementioned styles of leadership lacked success or had poor relationships with their men. Many actually achieved fine results, earned respect (although not devotion), and fully deserve the laurels bestowed by their grateful nations.

History also reveals a smaller number of commanders of such outstanding abilities that they earned immortality by employing the pure form of leadership defined by Howard: the gift of inducing total loyalty and devotion, and the ability to inspire, to motivate, and to enthuse. At the highest levels (after all, popular and rousing small-unit leaders are fairly numerous) these remarkable chieftains can be numbered on the fingers of one's hands: Alexander the Great, Caesar, Wolfe, Napoleon, Nelson, Lee, Jackson, Grant, Patton, and Rommel. Nelson is the only seaman.

Nelson's men gloried in their leader. Shortly before the Battle of Trafalgar, Capt. Thomas Fremantle wrote to his wife: "We are all busy scraping our ships' sides to new[ly] paint them in the [black and yellow] way Lord Nelson paints the *Victory*."[46] This puts one in mind of, say, the devotion felt by Napoleon's soldiers, who delighted in painting great Ns on buildings as they advanced across Europe; or of the troops of panzer leader Heinz Guderian, who proudly painted large Gs on their tanks and armored vehicles in a similar tribute to their beloved leader. Patton and Rommel inspired a variety of similar gestures of devotion.

Nelson realized that he served as a role model to his "darling children," as he called his officers and seamen.[47] They served in his "school," he told Lord Keith in 1800, much as he, Keith, and others had gained their own "professional zeal and fire" from St. Vincent.[48] Nelson was right to describe his influence on subordinates as a school—he certainly taught them far more about warfighting than they learned from anyone else—and he was particularly insightful to see that the greatest lessons were those that imparted professional zeal and combat aggression.

As a man of action and passion, he found these lessons easy to teach via that greatest of instructional tools: personal example. He had himself learned this way. Early in his career his "sea-daddy," Capt. William Locker, had inspired Nelson with accounts of how, during the Seven Years' War, Locker had boarded an enemy privateer by boldly leaping across from his own ship, the *Experiment*. His leg badly damaged in the fall, he nonetheless took the privateer and earned widespread praise for his courage. Nelson was spellbound by the heroic deed and thereafter modeled his own combat courage on Locker's.[49] Thus he knew well the importance of role models.

And in terms of inspiring courage and combat zeal through his own example, Nelson was a role model par excellence. In every engagement or battle, he demonstrated great physical courage and almost always led from the front. During the Battle of Cape St. Vincent on 14 February 1797, he called for a boarding party and with sword drawn jumped onto the stern of the Spanish ship *San Nicolas*. After fighting their way into that ship under heavy fire, Nelson and his men reached the quarterdeck and accepted the officers' surrender. Leaving a party to secure the ship, they then leapt aboard a second ship, the *San José*, taking that one as well. This remarkable act of heroism earned Nelson a tremendous reputation as a warrior and inspired his crews, who proudly described his unprecedented feat of capturing two ships by crossing over the first to attack the second as "Nelson's Patent Bridge for Boarding First-Rates."

Likewise, on 3 July 1797, he raided Cadiz with a bomb vessel protected by the boats of the fleet. As a rear admiral, he was expected to remain back on his flagship. Nelson ignored this expectation (and all precedent), and disobeyed his wife's worried insistence that he stop making assaults and boarding enemy vessels.[50] He led the Cadiz raid himself. Marvelously inspired, his men shouted "Follow the Admiral."[51] In the ferocious hand-to-hand fighting that ensued, the rear admiral was in such grave danger that his coxswain, John Sykes, twice saved his life, once stopping a sword blow meant for Nelson by interposing his own hand.[52] Nelson's willingness to share all dangers with his officers and seamen once again roused their own valor and left them with the lasting impression that he was the very model of commitment and gallantry they should emulate.

Leading from the front may be terribly inspiring, and consequently of immeasurable morale-boosting value, but it vastly increases a commander's likelihood of being wounded or killed. Later in 1797 at Santa Cruz, Nelson insisted on leading his men personally after the first attacks, under Troubridge, had failed. Nelson's courage could not compensate for a poor plan, fierce Spanish resistance, a lack of surprise, and adverse oceanic and weather conditions. His attack failed—badly—and cost him his right arm and the loss of many men. No one blamed him (except for himself), and everyone marveled at his daring. One seaman was so inspired that he tore a piece of lace from the sleeve of Nelson's amputated arm and thereafter kept it tucked in his clothing, later telling the admiral (who was visibly moved by this gesture of devotion) that he would keep the lace until his own death.[53] Even gruff St. Vincent, doubtless disappointed by the failure, felt deeply impressed by Nelson's fearlessness and promised to bow before his stump as soon as they met.[54] Years after Nelson's death at Trafalgar, and following fresh public revelations about his adultery with Emma Hamilton, St. Vincent fumed about his great protégé's disgraceful behavior. But he did have "animal courage," St. Vincent conceded.[55]

Nelson—like Napoleon, Lee, Patton, Rommel, and a few others—enjoyed the adrenaline-pumping thrill of battle almost as much as he detested the human misery it caused. After Troubridge's ship the *Culloden* ran aground during the Battle of the Nile, Nelson reassured him that he had done nothing wrong. Nelson merely regretted that his friend had to watch the squadron, "in the full tide of happiness," smashing up the French fleet, while Troubridge could take no part in the excitement and glory of the battle.[56]

One might see this phrase "the full tide of happiness" as a distasteful attitude toward combat, which participants usually consider the most frightful and chaotic experience of their lives. Yet Nelson, prematurely aged and exhausted by decades of intensive sea service, always felt reinvigorated and strengthened by battle and never shied away from its dangers. At the Battle of Copenhagen in 1801, he cheerfully remarked to Col. Sir William Stewart, who later wrote a lively account of the battle, that it was "warm work" which could end their lives at any second. "But mark you!," he told Stewart as bomb splinters flew around them on the quarterdeck of the *Elephant,* "I would not be elsewhere for thousands."[57]

The influence of Nelson's warrior spirit on his captains was immense. It gave them a model of leadership that they readily emulated. At Copenhagen, Capt. Edward Riou lamented that he had to withdraw his small frigate division from combat in obedience to an order from Sir Hyde Parker. Riou desperately wanted to be in the thick of the action, fighting courageously for his nation as he knew Nelson would be. With the depressed exclamation—"What will Nelson think of us?"—Riou gave orders for his frigates to bear out of the fight.[58] Thankfully other captains never got to see Parker's order and consequently battled on, each trying to fight with the tenacity and courage of their one-armed admiral.

Similarly, Nelson's second in command at Trafalgar, the talented and dependable Collingwood, led his column into battle with the hope of acting as courageously as Nelson (not that Collingwood ever lacked courage; he was a hero in his own right). When his column began pulling ahead of Nelson's, Collingwood, delighted that he would beat his friend into the frightful maelstrom, laughed to his captain: "Rotherham, what would Nelson give to be here?"[59]

The desire among Nelson's subordinates to match his tenacity, courage, and offensive ardor did not die with him. On 13 March 1811, one of his favorite "children," Captain Hoste, led a weak British squadron to a fine victory over a stronger Franco-Venetian squadron in the Adriatic Sea. As the two squadrons closed for battle, Hoste signaled to his squadron to remember Nelson. Nothing more needed to be added. Everyone knew what the signal meant.

Nelson's remarkable ability to inspire courage and combat aggression was not the only manifestation of his leadership skills. He also managed to simplify the direction of operations to their three basic elements: simple plans, clear communications, and operational and tactical flexibility. Through long periods of reflection on warfighting—to which he devoted his attention during spells of shore leave or routine naval patrols—he formed uncomplicated overall concepts for forthcoming naval operations. Everyone involved in these operations would need to understand fully his particular role as well as Nelson's overall intention. He therefore encouraged constant dialogue with and among his subordinate officers, having them dine with him on a regular basis so that he could give guidance in person and listen to other ideas or concerns.

Captain Berry recalled how, during the Nile campaign of 1798, Nelson repeatedly gathered his captains together on his flagship, the *Vanguard,* "where he would fully develop to them his own ideas of the different and best modes of attacks, and such plans as he proposed to execute upon falling in with the enemy, whatever their position or situation might be by day or by night."[60] A passionate student (and master) of naval tactics, Nelson patiently schooled his officers so well in a range of tactical options that by the time they engaged the French fleet, they were themselves capable of conducting superb maneuvers without the necessity of centralized control via the slow and awkward signal systems then available. The advantages of this schooling, Berry concluded, "were almost incalculable."[61]

Nelson also convinced his subordinates that he wholeheartedly trusted their judgment and abilities. They could and should, he reassured them, use their initiative and act as they thought best to meet his objectives despite any changing tactical circumstances.

Management (Humans)

One might naturally assume that Nelson was a poor administrator because of his impatient, excitable nature. Indeed, he probably would have proved inadequate if ever elevated to a senior Admiralty managerial position. St. Vincent certainly thought so. In November 1800, shortly before his appointment as first lord of the Admiralty, St. Vincent told the Admiralty's secretary, Sir Evan Nepean, that Nelson "cannot bear confinement to any object . . . and never can become an officer fit to be placed where I am."[62] This evaluation is probably correct; Nelson was a warrior, as unsuited to desk jobs as, say, the intrepid generals "Jeb" Stuart and Erwin Rommel would have been. These men were temperamentally suited to the cut-and-thrust of combat.

That is not to say warrior commanders made poor administrators. On the contrary, they knew full well that to win battles and campaigns, they had to keep their forces well supplied, which necessitated careful prioritizing of resources and time. Nelson may not have flourished in strategic management (planning and directing war at the national political-military level), but he did exceedingly well in operational management (planning, resourcing, and directing battles, campaigns, and blockades).

The greatest resource available to any commander, but perhaps the hardest to manage, is people. Nelson, who delighted in his relations with officers and seamen and genuinely cared for their well-being, always paid attention to what we now call Human Resource Management; that is, the training, development, and appointment of personnel in order to satisfy the aspirations and needs of individuals as well as of the greater group or organization.

Responsibility for the careers of hundreds of seamen on each of his ships was naturally too big a task for Nelson (or *any* commander) to oversee personally. He trusted his junior officers, petty officers, and mates. But he directly involved himself in the training and career progression of his officers, ensuring that they became skilled, professionally confident, and honorable gentlemen. As he wrote to one young boy about to be promoted to midshipman: "And as you from this day start in the world as a man, I trust that your future conduct in life will prove you both an Officer and a Gentleman. Recollect that you must be a Seaman to be an Officer, and also that you cannot be a good Officer without being a Gentleman."[63] He must continue to earn his captain's respect and protection, Nelson told the young fellow, and if he did so he could feel confident of his admiral's assistance as well.

Nelson's advice is revealing; it demonstrates how naval career advancement took place in an era when no regular structured training programs and few formal examinations (except for the King's Commission as lieutenant) existed. The Admiralty alone had the power to grant the King's Commission or confirm most on-the-spot acting promotions.[64] Yet, the patronage and mentoring of individual midshipmen and officers by their captains, and of captains and commodores by their admirals, provided the normal means of raising men through the ranks or getting them into roles or posts where their talents would shine. This patronage would naturally only occur if the officers demonstrated the zeal, professional skills, and personal attributes that the navy considered desirable. No one wanted to be involved in the promotion of *duds*.

Nelson, like most of his peers at this time, understood his own personnel management responsibilities to include:

1. Maintaining an awareness of subordinates' interests, strengths, and weaknesses. Nelson, sociable and chatty by nature, and ever watchful of his subordinates' actions, found this responsibility easy to accomplish. He sent a steady stream of letters conveying his impressions of his subordinates to St. Vincent and other influential persons. For example, in December 1803 he closed a letter to St. Vincent, in which he complained at length about the state of his fleet, with these words of praise for one deserving subordinate: "Sir Richard Bickerton is a very steady, good officer, and fully to be relied on."[65] This assessment is fair and accurate: Bickerton, a sensible, dependable, and seemingly unflappable man, provided a great deal of stability to Nelson's fleet.
2. Finding ways of using subordinates' strengths and minimizing their weaknesses. Nelson seldom reported an officer's weaknesses or inferior qualities to the Admiralty (unless they resulted in serious misconduct or dereliction of duty), preferring instead to work with his own team to help the officer improve his abilities in the area of deficiency. For example, like any good commander of the era, he often placed a seemingly cautious or reticent officer under the guidance of experienced, paternal, and assertive captains—or under his own watchful eye if he felt really concerned. These experienced warriors would impart assurance and strength and hopefully inspire the officer to act with greater confidence.

   Nelson's own mentoring of Don Domingos Xavier de Lima, the marquis de Niza, a young Portuguese admiral he came to like and trust during 1798 and 1799, is a fine example of Nelson's desire to help subordinates reach their full potential. During the sixteen months that Nelson commanded Niza, he gave him increasingly challenging roles as well as plenty of encouragement, close guidance, and thorough feedback. He thus helped transform the marquis from a barely competent seaman and poor admiral into a true sea-officer that he would praise as a "brother."
3. Ensuring that the persons most suited (either by inclination or experience) to certain functions were actually assigned to them. The admiral

noticed that certain officers were well suited to particular roles, such as commanding a frigate (as Sir Henry Blackwood and Parker did masterfully), serving as flag captain to a commodore or admiral (Hardy), commanding a line-of-battle ship (Berry), or leading a detached squadron (Troubridge). Respecting their unique talents, Nelson gave them greater responsibilities and opportunities for glory within those roles, but seldom sought to place them in other roles unless they indicated a desire to do so. Hardy, for example, was an exceptional captain, but an even better flag captain, and Nelson repeatedly sought to have Hardy for his own flagships.

4. Making clear that meritorious service had tangible career-related rewards (public praise as well as formal promotion or other advancement). Nelson never failed to praise enthusiastically and publicly—and, when possible, promote—any officer or seaman who distinguished himself through courageous acts, excellent performance of duties, or meritorious acts of selflessness. His praise was often profuse. In May 1796, for instance, he wrote to his commander in chief:

   > Much as I feel indebted to every Officer in the Squadron, yet I cannot omit the mention of the great support and assistance I ever receive from Captain Cockburn. He has been under my command near a year on this station; and I should feel myself guilty of neglect of duty, were I not to represent his zeal, ability, and courage, which shine conspicuous on every occasion which offers.[66]

It is worth noting Nelson's phrase "neglect of duty." He clearly understood that one of his key functions was to see worthy subordinates rewarded and hopefully promoted for excellent performance. Failure to bring their activities to the attention of his superiors, on the other hand, would be nothing short of professional negligence.

Unlike many of his peers (but like his mentor St. Vincent, and like Napoleon and George Patton), Nelson believed that promotion or decoration should immediately follow the pleasing deeds of subordinates. It was, he said, "absolutely necessary that merit should be rewarded on the moment" so that others would see that their commander's opinion of them actually mattered.[67] In order to increase the likelihood of the subordinates' further career advancement, he also routinely sent details of those pleasing deeds to his superiors with affirmations that the officers could and should be trusted with greater responsibilities or better assignments.

Interestingly, in March 1799 he gave his friend Troubridge a piece of advice that perfectly sums up Nelson's entire philosophy of human resource management. Always bear in mind, he said, "that speedy rewards and quick punishments are the foundation of good government."[68]

Nelson excelled at this unsophisticated but generally effective form of human resource management. As noted above, he delighted in his crews and became familiar and friendly with his officers, knowing them by name, career experience, and, perhaps most importantly, nature. There are too many examples for an inclusive analysis, but the case of Sir Edward Berry is so representative of Nelson's commitment to a promising subordinate that it may prove illustrative.

"Here comes that damned fool Berry," Nelson exclaimed on 13 October 1805, shortly before the Battle of Trafalgar, "*Now* we shall have a battle!"[69] Nelson was not disparaging Berry, who was a fine seaman, courageous warrior, and firm friend. Nelson was praising him for his remarkable fighting spirit and could not wait to glory in Berry's valor. He had first witnessed this back in 1796, when Berry's energy and courage convinced Nelson to entrust his ship, the *Captain,* to Berry while Nelson fought ashore on Corsica. For Berry's faithful care of the ship, Nelson (with his wife Fanny's urgings) had successfully recommended Berry's promotion to captain.[70]

Berry won further laurels during the Battle of Cape St. Vincent in 1797, when he joined Nelson in boarding the *San Nicolas.* In an act that reminded Nelson of his own patron Locker's daring jump from the *Experiment* decades earlier, Berry heroically threw himself from the *Captain* into the *San Nicolas*'s mizzen chains. After Nelson and the boarding party followed, they found Berry already in triumphant possession of the poop deck.[71] Nelson was so delighted that he not only profusely praised Berry to St. Vincent (who promptly promoted Berry to post captain), but also promised him the role of flag captain when next Nelson received a squadron.[72] Nelson kept his word, and Berry, as one of Nelson's famous "band of brothers," served in that capacity during Nelson's victory at the Nile.

Immediately after the battle, Nelson praised all his marvelous captains and crews in a letter to St. Vincent. Yet, in order to repay Berry's excellent service by hopefully advancing his career, Nelson singled him out for special praise: "The support and assistance I received from Captain Berry cannot be sufficiently expressed." Even though Nelson was wounded and had to be taken below, Berry remained in control of the ship's combat and proved successful in all his efforts. "Captain Berry was fully equal to the important service then going on."[73] St. Vincent and the Admiralty agreed, and Berry was knighted shortly afterwards. He then served as Nelson's captain on the *Foudroyant* while Nelson recuperated at Naples. Berry performed well under Nelson's guidance, playing a key role in the capture of the *Généreux* and *Guillaume Tell,* the two French ships that had escaped destruction in the Battle of the Nile.

Berry soon departed from Nelson's service, gaining command of the *Agamemnon,* in which he did nothing memorable. Away from Nelson's inspiration and guidance, Berry proved himself lacking in initiative and apparently incapable of assuming greater responsibilities. Without direction from a superior of wider capabilities, the brawler's career stalled. After the collapse of the Peace of Amiens in 1803, he was effectively "left on the shelf" without a command, his reputation as a ferocious scrapper his only consolation. In August 1805 he appealed directly to Admiral Nelson for assistance.[74] The admiral remembered his former subordinate's remarkable tenacity and courage and wanted it present in what Nelson expected would be a decisive battle. Thanks to his great patron, Berry duly received command again of the *Agamemnon.* The rest, as they say, is history.

Management (Logistics)

Nelson took care not only with the management of people, but also with material resources. Even as a young captain, he devoted himself to what we now call logistics. He took great care to ensure that through regular maintenance, his ships remained seaworthy and capable of effective warfighting; that the supply of essential water, provisions, stores, and munitions (totaling up to two hundred tons in a frigate) occurred smoothly and regularly; and that these were distributed fairly and economically to the crews. This task required careful planning and great attention to detail.

Nelson demonstrated both, and when he gained flag rank, he generally managed to keep his various ships and crews in very good shape. During 1798, 1799, and early 1800, he maintained continuous blockades of Alexandria and Malta and conducted successful naval and joint operations around Italy. His squadrons managed to perform these tasks and remain seaworthy (although sometimes only just) during all seasons even though they were scattered as far apart as Alexandria, Malta, and the Gulf of Genoa. Nelson's management of his British and Portuguese squadrons during this period was truly superb. Indeed, it is a pity that his effective managerial efforts have been overshadowed by his blossoming affair with Emma Hamilton and his growing involvement in Neapolitan government.

Even so, best efforts were sometimes not enough to overcome seriously adverse conditions, as his ship's condition back in August 1794 attests. After fifteen months of continuous and arduous escort, patrol, and combat duties around the northern Mediterranean, where all four seasons provided their own harsh punishments to ships and men, Nelson's *Agamemnon* was in a poor state. So was its crew. Nelson and many of his seamen were fighting ashore near Calvi, and they were suffering just as dreadfully as those on board the ship. "I am in the best health," Nelson told Lord Hood in August 1794 (probably exaggerating at least a little), "but every other Officer is scarcely able to crawl. . . . today it is as easy to keep a flock of wild geese together as our Seamen. We number ninety sick."[75] Supplies were almost always overdue, the weather was bad, exhaustion was taking its toll, and the "Calvi fever" was spreading wildly. Nelson finally received permission to remove his ship to Leghorn for a refit and to allow the crew to recover. After two weeks there, he reported to his wife that his crew's recovery was slow and few were ready for service again.[76] A month later he glumly told her that his ship's company was by no means recovered, yet they would serve and fulfill their responsibilities until the ship and crew simply couldn't continue.[77]

Nelson learned an important lesson from this and other unpleasant early experiences: one of an admiral's most crucial tasks was the effective provisioning of ships. When he met Sir John Jervis for the first time on 15 January 1796, soon after the old admiral gained command of the Mediterranean, Nelson was delighted to find a man who made the condition of ships and the welfare of crews his highest priorities. Improvements began almost immediately, to the relief of the whole fleet.

Seven years later in 1803, Nelson himself gained supreme command of the Mediterranean, thus inheriting one of the navy's greatest logistical challenges: how to keep a whole fleet well supplied in a generally hostile environment with few friendly, and even fewer adequate, supply and victualling ports. Arriving off Toulon in July 1803 to throw a blockading screen around the French fleet within the port, Nelson stated: "My first object must be to keep the French in check and, if they put to sea, to have force enough with me to annihilate them."[78] The word "annihilate" is so dramatic that it distracts attention from what, by comparison, seems unimportant: the words "to have force enough." Yet it is clear that these words describe one of Nelson's key aims. He knew blockade duties would drag on for at least several months and possibly, "God forbid," for over a year, but at the end of that exhausting and logistically difficult period he expected his force would still be fully able and ready at a moment's notice to chase, engage, fight, and win.

This commitment would not be easy. After reaching Toulon he wrote to St. Vincent on 8 July 1803 to report that although his ships looked fine above the waterline, their "bottoms" were not in great shape.[79] They were also all undermanned (by as many as a hundred in each ship of the line) and had worn sails and rigging.[80] St. Vincent informed him, effectively, that the Admiralty would supply what it could, which was not much, and that Nelson would simply have to make things last through strict economy.[81] This message gave neither sender nor recipient any pleasure.

Even worse, Nelson now faced a dilemma that his predecessors had not encountered in the late 1790s: he had no adequate harbor that he could use as a command, maintenance, and victualling base. Genoa, Leghorn, and Corsica were back in French hands; Minorca had reverted to Spain; Gibraltar and Malta, although British (or British controlled), were too far from Toulon, as were Naples and Palermo, which were now neutral anyway.

Nelson addressed this problem with typical ingenuity. He decided that his fleet as a whole would remain continually at sea and (except for individual ships of the line needing urgent repairs) never go into port; that is, until the French attempted a break-out from Toulon and were defeated in the act.[82] The fleet, he told St. Vincent, was particularly ready and willing to "give the French a thrashing."[83]

To keep his fleet capable of meting out this punishment would require the most careful and complete management. Nelson rose to the challenge. He organized storeships to bring supplies from Gibraltar and Malta, as well as from Sardinia and the Kingdom of the Two Sicilies, whose rulers, although technically neutral, remained in Nelson's debt and were willing to offer whatever surreptitious assistance they could. Even several Spanish ports agreed to sell meat and vegetables to the fleet, ignoring their nation's neutrality agreements with France. Nelson spent long hours each day with his secretaries, the two Mr. Scotts, one of whom (the often-unwell chaplain) he called "Dr." Scott. They drafted letters of great diplomacy to local and regional authorities asking for the right to purchase essential materials and foodstuffs.[84]

The admiral worked hard and carefully to ensure that his fleet remained healthy. He devoted much time to the correct victualling of his ships, sending a steady stream of requests to the Admiralty (particularly for firewood, coal, and candles) and orders to the victualling officers at Gibraltar as well as to individual captains within his fleet. Fresh (not old and salted) meat and vegetables were essential to keep his crews healthy and safe from scurvy, and these he insisted on. In one such order—dated 8 March 1804, after his fleet had been at sea for almost a year and had endured a dreadful winter—he told Capt. James Moubray of the *Active* to purchase

> fifty head of good sheep for the use of the sick on board the different Ships, with a sufficient quantity of corn and fodder to last them a month [that is, to keep the sheep alive until they were needed to provide *fresh* meat]; and if it is possible to procure thirty thousand good oranges for the Fleet, with onions, or any other vegetables . . . you will also purchase them, together with as many live bullocks for the Ship's Companies as you can conveniently stow, with fodder to last them during your passage.[85]

This does not mean that all crews in his fleet escaped having to eat salted meat. On 25 April 1804, seaman John Booth of the *Amazon* wrote to his wife: "We are all tir'd of being at sea on salt provisions for we have not been in an English port this 6 months and don't expect of 5 or 6 more to be in one."[86] Likewise, in March 1805, when it appeared that the French had slipped out of Toulon and headed for Alexandria, Nelson sailed there with great haste. The voyage there and back to Sardinia left his ships "reduced to a small quantity of salt provisions, and [even] less stinking water."[87] Urgent victualling immediately took place. Despite such times of difficulty, however, good health and reasonable morale remained the norm, much to Nelson's credit.

One thing that Nelson's fleet needed regularly and urgently was water, which the storeships could not bring in sufficient quantities. He therefore chose the Maddalena Islands off the northeast tip of Sardinia as his fleet's principal watering point. It was also an ideal spot for replenishing supplies of timber and firewood.[88] He sought good relations with local inhabitants, using the linguistic and diplomatic skills of Dr. Scott to facilitate clear communications. In return for what Nelson called their excellent assistance, he bestowed gifts upon the local parish church, including expensive engraved silverware.[89] The Maddalena Islands were a fine choice: "The Fleet being very much in need of water," he told Hugh Elliot, "I have taken the opportunity of the moonlight nights to come here, in order to obtain it, and some refreshments for our crews . . . our health and good humour is perfection."[90]

The ships themselves were a different story. They had been in poor repair even before Nelson's arrival in the Mediterranean, and by December 1803, after six months or so of continuous blockade duties, Nelson complained to Alexander Davison: "My crazy Fleet are getting in a very indifferent state, and others will soon follow. The finest Ships in the Service will soon be destroyed."[91] He was well aware, he told Davison, that his ships could escape the worst weather if they sought shelter at Malta, but that would create the grave risk that the French fleet would slip out unnoticed. If that was to be prevented, he concluded, he had no choice but to keep his fleet at sea throughout winter. He told St. Vincent at this time that sending his fleet to Malta to escape the worst weather would be so disadvantageous to the blockade that he might as well send them to Spithead in England.[92] Sending individual ships to Malta for urgent maintenance took them out of service for two months, prompting Nelson to curse the island as "useless" to him.[93]

He therefore devoted a lot of time to surveying at regular intervals the state of the fleet as well as the amount and quality of stores held in the military warehouses at Gibraltar and Malta; ordering the right quantities of rope, canvas, and timber needed to keep ships functional;[94] checking that the exact quantities arrived (and were not reduced at all by dishonest transport captains who sold some for personal gain);[95] and rostering the removal of vessels for maintenance so that few, and only those in the worst condition, were absent from blockade duties at any one time.

All these administrative duties—which he worked at with surprising care and patience despite his frequent seasickness and constant poor health—certainly paid off. For almost two years his fleet continued to watch the French Mediterranean fleet, keeping it under close guard despite the exhausting nature of these duties. Yet, when the French finally did slip out of harbor, Nelson's own fleet was still in good fighting condition, its crews healthy and keen for battle and its ships seaworthy and ready for whatever pursuits or voyages were required. And they did indeed prove capable of a long voyage, relentlessly pursuing the French fleet across the Atlantic and back on a three-month, seven-thousand-mile-round chase.

So effective were Nelson's administrative efforts in this period that the popular naval writer C. S. Forester wrote: "The achievement stands as nearly the greatest, and by far the least known, of all Nelson's claims to fame."[96] In a similar vein, although with far more words, the great sea power theorist, Alfred Thayer Mahan, explained that

> Nelson's command during this period [was] a triumph of naval administration and provision. It does not necessarily follow that an officer of distinguished ability for handling a force in the face of the enemy will possess also the faculty which foresees and provides for the many contingencies, upon which depend the constant efficiency and readiness of a great organized body; though both qualities are doubtless essential to constitute a great general officer. For twenty-two months Nelson's fleet never went into a port, other than an open roadstead on a neutral coast, destitute of supplies; at the end of that time, when the need arose to pursue an enemy for four thousand miles, it was found massed, and in all respects perfectly prepared for so distant and sudden a call.[97]

Nelson's awareness of the importance of correct victualling and timely supply, coupled with his willingness to devote ample time to these administrative duties, remained with him even during his nerve-racking preparations for his final showdown with the French. Several times during the weeks leading up to the cataclysmic Battle of Trafalgar in October 1805, he reduced his combat strength by diverting ships to Gibraltar on watering and victualling duties.[98] On one occasion he sent off six powerful ships of the line, so important did he consider his logistical responsibilities.[99]

Conclusions

My experiences as a university defense and strategy lecturer convince me that something critical is now wrong with, or missing from, the teaching of command studies. In today's learning culture, sophisticated and jargon-filled theoretical constructions are more popular as teaching tools than the study of past and current human experiences. As a result, *command* has lost its true meaning and become a system or method that can be learned (supposedly; I deny it) away from the environment in which it is practiced. Even in many defense studies departments, command is now taught via the study of management paradigms that are purportedly as applicable to business and the boardroom as they are to warfare and the battlefield.

A study of Nelson's command style, however, goes some way to dispelling this nonsense. It reveals that command in war is a remarkably complex, almost indescribable, deeply human relationship between those who lead and those who follow in circumstances usually dangerous, chaotic, and frightening. Command in war requires a leader to do certain things never required of the most ambitious and competitive businessman or -woman: to be courageous and decisive in the face of mortal peril; to inspire subordinates to act with courage and aggression; to carry out the violence necessary to ensure victory; to provide restraints on that violence in order to meet decent standards of morality, justice, and legality; to assume direct responsibility for the lives and health of all involved.

Nelson willingly shouldered these heavy responsibilities and mastered them so fully that he, alongside only Napoleon, stands head and shoulders above all other commanders of the era (metaphor is a wondrous thing; it allows two noticeably short men to tower over others). At the operational level, Nelson was a commander, leader, and administrator of outstanding aptitude. His officers and seamen virtually worshipped him and consequently performed wonders in his service. Under his guidance and loving attention they developed a confidence, morale, and fighting ability without equal. Even seamen who had never met or seen him—and many in his fleets had not—felt somehow connected to him. They knew him from comrades who had served on ships under his command and from the legend that had emerged in his own lifetime. They knew about his wounds, his mutilated body, his self-sacrifice, his patriotism, his tactical brilliance, and perhaps most important in their eyes, his care, respect, and affection for them. Many—even tough and battle-hardened old warriors—wept openly after learning of their hero's untimely death.

An apt quotation to close this chapter describes the loving devotion he inspired:

> The final incident of Lord Nelson's funeral, found by many spectators the most impressive, was undisciplined and unrehearsed. It had been set down that the men of the *Victory* were to furl the shot-rent colours which they had borne in the procession and lay them upon the coffin; but when the moment came, they seized upon the ensign, largest of the *Victory*'s three flags, and tearing a great piece off it, quickly managed so that every man transferred to his bosom a memorial of his great and favorite commander.[100]

Such spontaneous acts reflect the genuine desire of mourners not to let death rob them of all connection to their beloved friend, guide, and champion. They are the truest mark of a man's greatness.

4

Nelson's Warfighting Style and Maneuver Warfare

FOR THE last twenty years or so, many who study warfare have come increasingly to see the warfighting style called Maneuver Warfare as the apex of military theory. The United States Marine Corps, which has formally embraced this style as doctrine, succinctly defines it as

> a warfighting philosophy that seeks to shatter the enemy's cohesion through a variety of rapid, focused, and unexpected actions which create a turbulent and rapidly deteriorating situation with which the enemy cannot cope.[1]

Proponents of this style call themselves maneuverists. They argue that adoption of the package of components integral to Maneuver Warfare will greatly enhance combat effectiveness. By adopting it, smaller maneuverist forces can even engage and defeat far larger forces that adhere to less sophisticated doctrines or behavioral patterns. Maneuverists claim that the components in their package—reconnaissance pull, the application of strength against weakness, decentralized command, focal points of effort, maneuver, tempo, and the aim of achieving victory by collapsing the enemy's cohesion and morale—are not in themselves new to warfighting. They are the long-established habits of successful commanders.

Maneuverists also argue that the combination of all these components occurred in many battles or campaigns throughout the ages and brought dazzling results (both claims are exaggerated, according to other sources).[2] Yet maneuverists provide examples mainly from the twentieth century and, more important, *mostly from land warfare*. The two standard works on Maneuver Warfare, products of the early 1990s when this warfighting style soared in popularity after its claimed use in the Persian Gulf War, provide few examples from, or references to, the rich treasure chest of naval history.[3]

It is the same with the articles on Maneuver Warfare now appearing in military journals. These articles focus overwhelmingly on land campaigns and battles. The exceptions tend to demonstrate that certain airpower and joint campaigns—such as those conducted by Hitler's Wehrmacht or the Israeli Defense Forces—also reveal the prowess of well-applied Maneuver Warfare. Far fewer articles on sea power and Maneuver Warfare exist. Readers seeking to analyze Maneuver Warfare's applicability to combat at sea can be forgiven for noticing the relative lack of scholarly interest in this theme and thinking that Maneuver Warfare must have limited applicability at sea.

One can, however, easily find many fine examples of what is now called Maneuver Warfare in sea power's long history. This chapter draws from one such example—splendidly manifest in the person of Nelson, Britain's greatest fighting seaman—to demonstrate that those who study maneuver need not fear turning their attention occasionally from land battles toward those fought at sea and may indeed be greatly enriched by doing so.

While being mindful to avoid anachronism (Maneuver Warfare's conceptual framework, after all, is very recent), this chapter shows that Lord Nelson's warfighting style closely *resembles* the modern Maneuver Warfare paradigm. He wasn't fighting according to *any* paradigm, of course, much less one that dates from almost two hundred years after his death. He understood naval tactics and battle according to the norms and behavioral patterns of his own era and continuously experimented and tested ideas, rejecting some, keeping others.

The chapter naturally makes no claim that Nelson's warfighting style was entirely unique among sea warriors or that he contributed disproportionately to conceptual or doctrinal developments in tactics or operational art. Even a cursory glance at the careers of John Paul Jones, Edward Hawke, Sydney Smith, and William Hoste (one of Nelson's protégés), to mention but a few, reveals that their names fit almost as aptly as Horatio Nelson's alongside Napoleon Bonaparte's, Erwin Rommel's, and George S. Patton's in studies of effective maneuverists. Yet Nelson makes an ideal focus for a case study of Maneuver Warfare at sea. Extant sources pertaining to his fascinating life are unusually abundant and reveal that he raised the art of war at sea to unsurpassed heights, all the while perfecting the highly maneuverist warfighting style that gave him victory in several of naval history's grandest battles.

Reconnaissance Pull

Maneuverists place much emphasis on what they term *reconnaissance pull*, by which they mean that attacks should move in directions identified by forward reconnaissance units, not by commanders in the rear who want to push their forces forward along preselected routes. Ideally, reconnaissance units should not only find the enemy, but also probe for undefended or lightly defended gaps in the enemy line that lead to the enemy rear.[4] Then the whole force should, upon orders from forward unit commanders, follow the *pull* of the reconnaissance units and stream through the gaps to achieve one or more penetrations and hopefully a breakthrough.

The aim of reconnaissance pull is to transfer some decision-making authority to forward commanders, who purportedly have a clearer and more immediate picture of enemy strengths, dispositions, and morale than senior commanders at the rear. It is also to ensure that one's strength is not directed at opposing strength, but at weakness.[5] This concept of avoiding or bypassing points of strong resistance to attack critical weaknesses, a central tenet of contemporary Maneuver Warfare, emerged as a popular theme from the writings of British military commentator Sir Basil Liddell Hart in the years immediately before and after World War II. It represents a clear departure from traditional Clausewitzian military thought. Writing in the 1820s, Carl von Clausewitz, the great Prussian philosopher of warfare, had explicitly advised that the most effective target for any blows is found where the enemy's mass is concentrated most densely.[6] Clausewitz was not advocating inexpert, bloody frontal assaults against this concentrated mass, of course; merely that a successful outcome ultimately rests upon the destruction or neutralization of that mass. Logically, then, the primary aim of combat is to attain that destruction or neutralization through aggressive and highly skillful attacks against the mass.

Nonetheless, following Liddell Hart's advice and ignoring Clausewitz's, maneuverists now argue that it makes more sense to avoid attacking mass and, by following the tactics of surfaces and gaps, to maneuver around strength in order to attack those shaky and weak parts of the enemy upon which he is critically reliant. Doing so, they say, offers no promise of bloodless combat, "but it does offer less bloody war than a head-on bash directly into the enemy's strength."[7]

This maneuverist emphasis on reconnaissance pull—along with information and decisions flowing upward as well as downward in the chain of authority, and with the avoidance of strength-on-strength clashes in favor of attacks on enemy vulnerabilities—will doubtless create a happy resonance in the minds of those who have studied Lord Nelson's fighting style. They will see many similarities between what he and other successful mariners were doing and what is now extolled as "best practice."

Nelson always relied heavily on his reconnaissance elements, and he insisted on maintaining the greatest degree of tactical and operational flexibility so that he could respond immediately to circumstances revealed by those elements. This is not to suggest that the conduct of reconnaissance at sea is the same as it is on land. At sea, opposing fleets or squadrons spend most of their time out of direct contact with each other, whereas opposing land forces are routinely in contact or close proximity, even when not actually in combat. This geographical anomaly creates different battle space dynamics and places different obligations on reconnaissance assets. At sea, far more effort is devoted to finding the enemy, whose location and direction are constantly changing (unless the enemy's vessels are riding at anchor), than on land, where the location of the enemy is usually known, even if his strength, disposition, and plans are not. Yet this does not negate the fact that fleets have always devoted as much effort to attaining battle space awareness as have army formations. And Nelson, like other great mariners, routinely allowed the flow of intelligence from his reconnaissance vessels to shape his campaigns and determine his tactics.

Since the mid-1700s, frigates—warships with a single gun-deck that functioned as scouts and convoy escort vessels since their speed was several knots faster than larger ships of the line—bore the lion's share of all naval reconnaissance responsibilities.[8] Nimble yet well armed for their size, frigates served as the eyes of fleets and squadrons: trying to locate the enemy; establishing his strength, dispositions, and direction of movement; and then reporting that information with all speed to larger fleet units. Nelson himself, although not uniquely, referred to frigates as his "eyes" at sea. "I am most exceedingly anxious for more *eyes,*" he complained to the secretary of state for war shortly before the Battle of Trafalgar in October 1805, "and hope the Admiralty are hastening them to me. The last [enemy] Fleet was lost to me for want of Frigates; God forbid this should [again occur]."[9] A lack of "eyes" had long plagued Nelson's efforts to bring the enemy to battle in advantageous circumstances. A full seven years before Trafalgar, but only a week after his stunning victory at the Battle of the Nile, Nelson had written to Earl Spencer: "My Lord, Was I to die this moment, 'Want of Frigates' would be found stamped on my heart. . . . [I] am suffering for want of them."[10]

Nelson used his reconnaissance elements to draw him into contact with the enemy and also relied on them to help shape particular tactics for employment once contact had been made. This practice did not involve frigates only, but any other vessels—including larger warships—which could probe the enemy line for weaknesses to exploit. Although the maneuverist concept of surfaces and gaps initially seems more applicable to linear formations in land battles, it does have relevance at sea, even if in a slightly different form. A line of warships (fleets or squadrons in the age of sail formed into lines before joining battle) still had points of strength and weakness. These were dictated not only by the positions of certain warships in the line, but also by tides, winds, currents, and the shape and proximity of nearby coastlines.

Early in the afternoon of 1 August 1798, for instance, the masthead lookout in one of Nelson's scouting ships sighted Napoleon Bonaparte's powerful fleet anchored in compact order of battle in Aboukir Bay, Egypt. This sighting ended Nelson's anxious two-month hunt throughout virtually the entire Mediterranean Sea. By the time Nelson's fleet closed on the French and formed its own line of battle, darkness was fast approaching. As already noted, few naval battles had ever occurred at night (and none of this scale), and the French admiral, Brueys, assumed that Nelson would attack in the morning rather than fight in darkness, which could cause friendly fire incidents and groundings on unseen shoals.

Brueys had long anticipated an attack and had deployed his vessels, slightly stronger than Nelson's in overall strength and firepower, with the possibility of a defensive battle in mind. He anchored his ships close to the shore in line of battle, with the head of the line close to shoals and a battery on an islet and with gaps of perhaps 160 yards between each ship. Given common winds, the shape of the bay, and the shoals near the head of the line, Brueys presumed that the British would enter the bay in a traditional line of battle, sail from the tail of his own line toward the van (the ships at the head), and bring his vessels under fire in sequence. With this in mind, Brueys had placed his strongest ships in the rear and center, believing that Nelson's ships would face devastating fire as they attempted to pass alongside the French ships toward the van.

Nelson was unorthodox, however, and had an unusually quick and agile mind. "Without much previous preparation or plan," one of his closest friends recalled, he had "the faculty of discovering advantages as they arise, and the good judgment to turn them to his use."[11] Indeed, Nelson quickly identified the major gap in Brueys's surfaces (to use maneuverist parlance), the short stretch of water between the *Guerrier,* a seventy-four-gun ship at the head of the French line, and the nearest shoal. The French ships rode on single anchors, Nelson had noticed, which logically meant that if there was room for the ships to swing on single anchors, there should be room for some of his own ships to pass around the head of the French line, hard up against the shoals, and to sail down the landward side.[12]

There, Nelson presumed, lay the enemy's critical vulnerability (to use maneuverist jargon again): unmanned or unprepared guns, leaving no real means of opposing an attack from the port (landward) side. Expecting to face the British sailing along the starboard (outward) side of his ships, Brueys would naturally have all his starboard guns and crews ready and waiting. But, not anticipating any attack from inshore, his port guns would probably be unprepared for action. This would allow British ships sailing along the port side of the French to fire at will while receiving, at least initially, relatively little return fire. It would also allow them to *double* (attack from both sides simultaneously) several ships in the van.

There is still much debate among Nelsonians about whether Nelson specifically ordered one or more of his captains to squeeze through the gap between Brueys's van and the shoals, or whether they made the attempt on their own initiative while aware that the daring maneuver matched their commander's intent and would, therefore, meet with his approval.[13] The current writer is persuaded by the evidence that Nelson initiated the bold move by hailing Capt. Samuel Hood of the *Zealous* and asking whether his ship could pass inshore of the French.[14] This initiated a friendly race between the *Zealous* and Capt. Thomas Foley's *Goliath* as they jockeyed to be first to get inshore of the French in order to move along the port side that would surely be ill defended.

Yet even if this view is wrong, and the initiative came from Foley (as some writers insist), the move is still a splendid early example of what we now identify as the Maneuver Warfare concepts of surfaces and gaps and the application of strength against weakness. It is also a fine example of another important maneuverist concept to be explained below: a form of decentralized command called Directive Control. And even if Foley initiated the move, Nelson also deserves credit. During the frustrating two-month chase through the Mediterranean, he had regularly assembled his captains and other senior officers on his flagship and explained to them his intentions. He also gave them detailed tactical suggestions for attaining his intentions regardless of whatever circumstances prevailed when the enemy was located. Equally important, he encouraged them never to fear using initiative by acting without orders as circumstances demanded. As one writer observed: "It was Foley's genius that he spotted the weakness. It was Nelson's genius that his captain felt secure enough in the admiral's trust to exploit the opportunity."[15]

Four of Nelson's ships were able to pass inshore of the French line, and a fifth passed between the first and second ships to join them. The rest completed the envelopment by sailing down the outer side of the enemy, from the van toward the rear. Thus, with British ships engaging the French van and center first and moving steadily down the line toward the rear, and with the wind behind them, they were able to attack the weakest French ships while the strong French rear was unable to come to their comrades' rescue.

This is a fine historical example of the maneuverist concept of pitting strength against weakness. And it worked marvelously. As Nelson's ships came up, they anchored in succession next to French ships, so that thirteen British were soon engaging eight French, and from positions of great advantage. The concentrated firepower of the British proved overwhelming, especially given that their rate of fire almost doubled that of the less expert French, forcing the French ships to strike their colors (a sign of surrender) one by one. The British then turned their attention to the French ships in the rear. Shortly after dawn the last guns stopped firing to reveal a British victory of singular completeness. Indeed, it was the most devastating naval victory of the eighteenth century. For no ships lost of their own, the British had virtually annihilated the French fleet. Six ships of the line had struck their colors, Brueys's flagship had sunk after a huge explosion ripped it apart, and four other ships were grounded with heavy damage on the beach. Only two ships of the line escaped (to be recaptured at a later date anyway). Casualty figures were equally one-sided: two hundred British and seventeen hundred French perished.

The Battle of the Nile is a superb early example of the maneuverist concepts of surfaces and gaps and the application of strength against weakness, but it is only one of many examples to be found in Nelson's illustrious career. Before he attained flag rank and commanded his own squadrons and fleets, he had already demonstrated on several occasions that he understood these ideas (not knowing, of course, that they would one day be extolled in doctrine as *principles* and *best practice*). Even when captain of a single ship within fleet units, he skillfully attacked far stronger opponents by avoiding their firepower and targeting their weaknesses. Off Genoa on 13 March 1795, for example, Nelson pulled his sixty-four-gun *Agamemnon* out of the British line (following Captain Fremantle in a smaller vessel, *Inconstant,* which had to break off pursuit) and sailed straight for the eighty-four-gun *Ça Ira*. This powerful French ship was being towed by a frigate after suffering damage to its masts during a collision with *Jean Bart,* another French ship of seventy-four guns.

Supporting *Ça Ira* was the huge 120-gun *Sans Culotte* (later renamed *L'Orient* and made famous as Brueys's flagship at the Nile) and the *Jean Bart*. Nelson was massively outgunned and would have suffered quick destruction if he fell under the enemy's fire. Undaunted, he sought and immediately found a chink in the enemy's armor: the stern of *Ça Ira;* the eighty-four, being towed, was less maneuverable than usual. Like all ships of the line, *Ça Ira*'s stern had few guns and was highly vulnerable to attack. With skill and courage (he later told his wife that "a brave man runs no more risk than a coward"),[16] Nelson tacked the *Agamemnon* up under the rear of the heavily armed Frenchman. He thus gained safety from the other French ships, and, at almost point-blank range, fired double-loaded broadsides again and again into *Ça Ira*'s vulnerable stern. For almost two hours he maneuvered back and forth behind

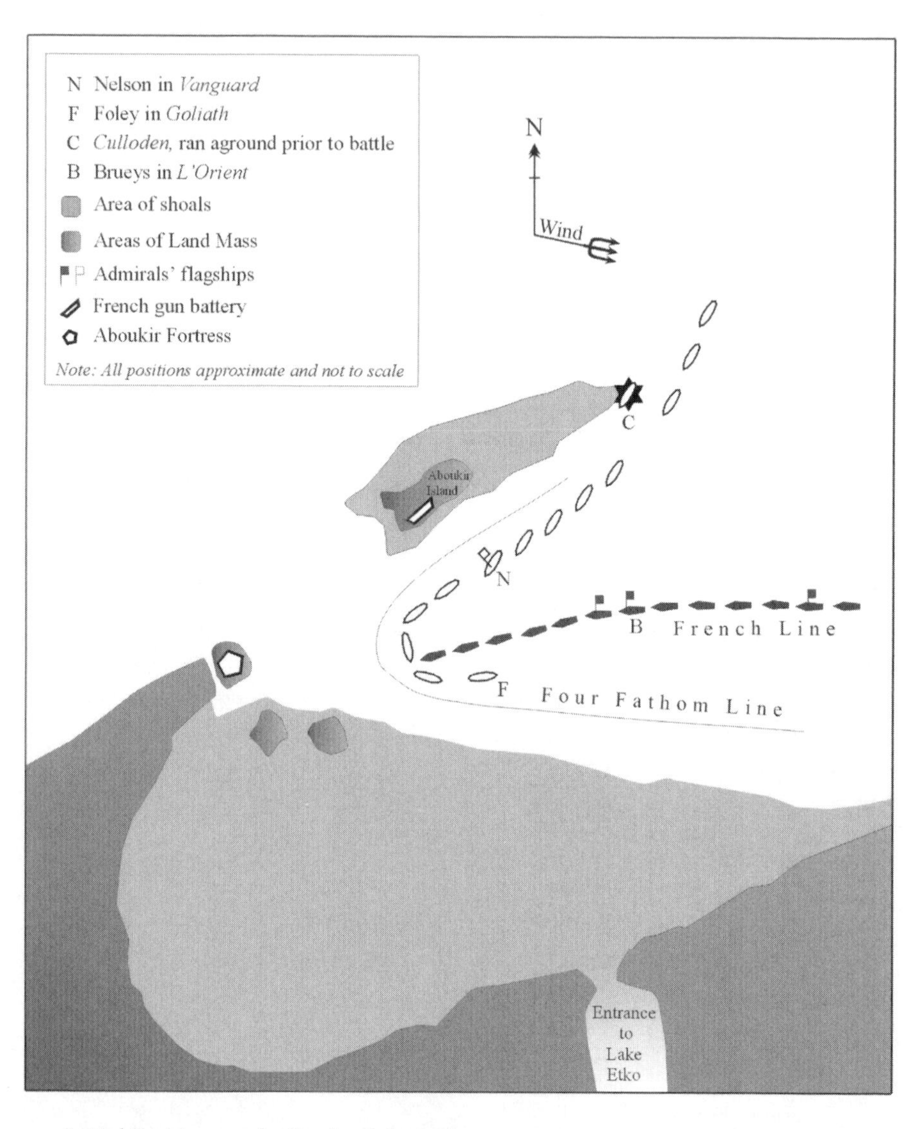

Initial Positions at the Battle of the Nile

Ça Ira, which he continuously raked while never allowing it to turn its own broadside against him. When *Agamemnon* finally broke off the engagement, with only seven men wounded, 110 Frenchmen lay dead and dying.[17] The Frenchmen's ship was, Nelson said, "a perfect wreck."[18]

Directive Control

As noted, Nelson's command style before and during the Battle of the Nile is a nice early example of that aspect of the Maneuver Warfare concept now called Directive Control. Actually, maneuverists use various terms synonymously to denote this form of decentralized command: Directive Control, Mission Tactics, Mission Command, and even the original German phrase, *Auftragstaktik* (task-focused tactics). Put simply—Directive Control means teaching subordinates mastery of their individual jobs and collective tactics, trusting them to act responsibly and with initiative, telling them *what result is intended* (usually defined as the condition or position you want the enemy to be in after the engagement), then leaving them with the freedom and confidence to determine how best to attain it.[19] In other words, commanders need to convey their *intent,* telling their trusted subordinates what the commanders want achieved, but not, at least in too much detail, the method the subordinates must employ to do it. General Patton, the warrior Nelson resembles most (for his vanity and craving for fame as well as his dazzling flair for battle), expressed it thus: "Never tell people *how* to do things. Tell them *what* to do and they will surprise you with their ingenuity."[20]

Even a superficial reading of Nelson's thirty-five-year naval career reveals him as one of military history's greatest and most consistent proponents of what we now call Directive Control.[21] As a young officer learning his trade, he was keen to study what had brought success in previous generations to the giants of British sea power, Hawke and Richard, First Earl Howe.[22] Hawke's victory at Quiberon Bay in 1759 was a splendid example of maneuverist warfighting, despite the limitations of eighteenth-century naval technology and tactics. It demonstrated to young officers like Nelson just how much could be accomplished with the minimum of direct command and control when subordinate captains were thoroughly familiar with their commander's intent, schooled in appropriate tactics, inspired by the commander's example, and trusted by him to use their initiative. Hawke and Nelson actually had a connection; Captain Locker, one of Nelson's important early mentors, had served with Hawke and come under his influence during his period of famous activity.

Shortly before his own great victory at the Nile, Nelson also had the good fortune of serving under Admiral Jervis, who demanded very high standards from his captains but, by training and trusting them, achieved excellent results. On 1 December 1796, Jervis entrusted Nelson with what Jervis knew would be a difficult mission: the evacuation of the British forces that were stranded at Porto Ferraio, Elba, following Corsica's evacuation. Jervis's written order is a model of its kind. He stated his intent clearly but simply and then informed Nelson that he should feel at liberty to accomplish the task as he saw best. "Having experienced the most important effects from your enterprise and ability," said Jervis, "upon various occasions since I have had the honour to command the Mediterranean, *I leave entirely to your judgement* the time and manner of carrying this critical and arduous service into execution."[23] Needless to say, Nelson lived up to Jervis's expectations.

Jervis's respect for Nelson bore sweet fruit again on 14 February 1797, when the admiral gained victory in the Battle of Cape St. Vincent. This battle firmly proved the value of trust—both *of* and *from* one's subordinates—and the clear communication of expectations, expressed not only in terms of the results that are wanted, but also of the skills, commitment, and initiative that each person must possess.

Jervis's fleet encountered a greatly superior Spanish force off Cape St. Vincent in Portugal, and the admiral immediately ordered its attack, aware that Spanish vessels were of adequate quality but that their crews and commanders were not. Jervis may have trusted his subordinates and wanted them to use initiative; but he was nonetheless of an older generation, was accustomed to exercising authority through centralized command, and therefore liked to control his engagements and battles by issuing signals from the flagship.[24] Nelson recognized this, of course, but also knew that the spontaneous circumstances of battle often rendered centralized command ineffective.

During the action off Cape St. Vincent, Nelson, then a commodore on board the *Captain,* watched with mounting frustration and then outright

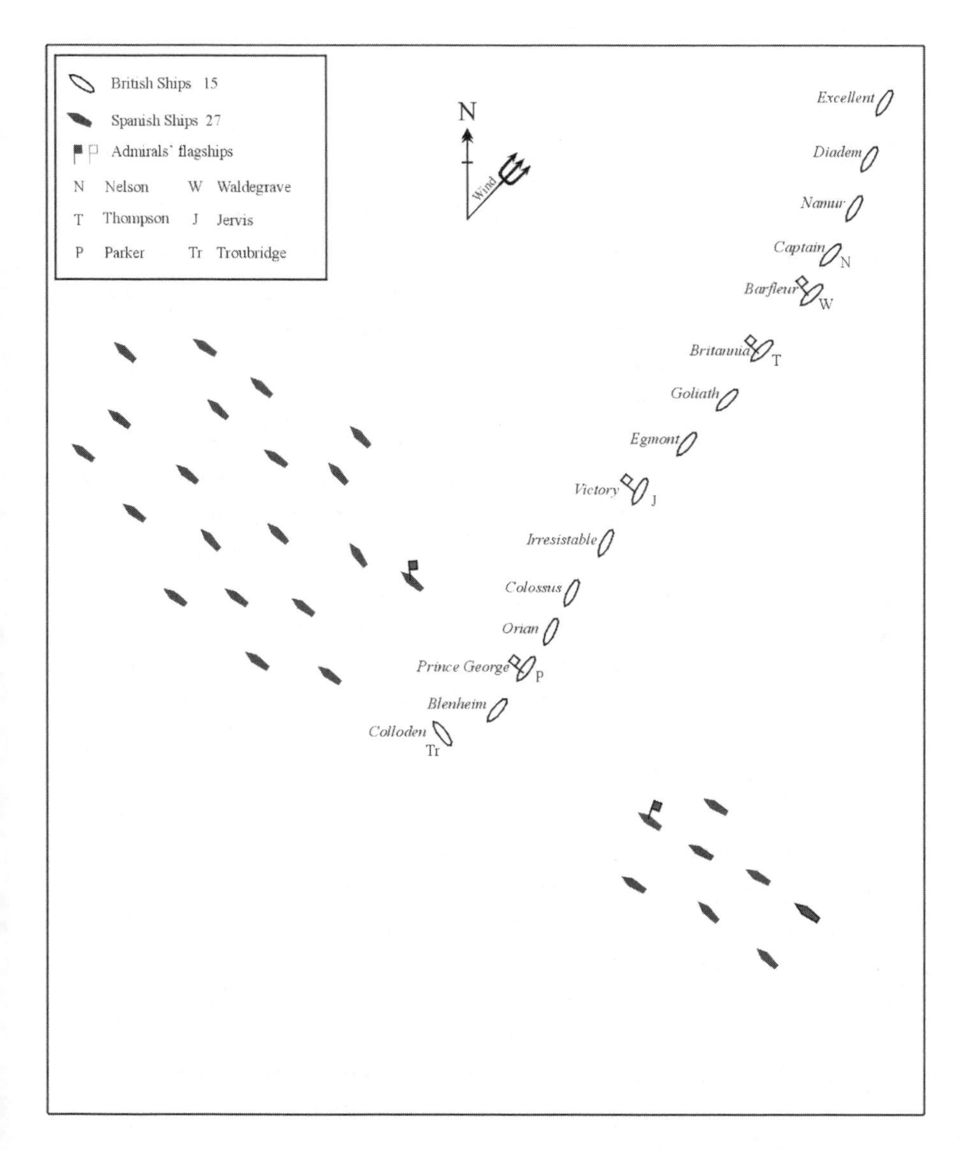

Initial Maneuver at Cape St. Vincent

angst as circumstances developed beyond Jervis's ability to control them with traditional maneuvers and communications by signals. Nelson saw a fleeting opportunity to cut off several ships in the Spanish van that seemed likely to escape due to slow and complex British maneuvers ordered by Jervis. On his own initiative, but still with Jervis's overall battle objectives in mind, he spontaneously wore the *Captain* out of the British line of battle, passed back through it, and attacked the Spanish van.[25] Even though Nelson knew that Spanish ships were poorly manned and incapable of reaching the British rate of gunfire (and he had long favored regular tactical drills to make his crews as fast and efficient as possible), his action was still highly courageous.[26] It occurred without approval from Jervis and also directly violated the *Fighting Instructions* that had governed British naval tactics for generations.

Yet even the conservative old admiral saw its genius and with the same signal system that had almost allowed his enemy to escape, promptly ordered Nelson's reinforcement (a marvelous early example of reconnaissance pull). Nelson's actions secured the victory and earned Jervis's profound gratitude. Sir Robert Calder, the cautious flag captain, later complained to Jervis that Nelson's action was unauthorized, to which the admiral replied: "It certainly was so . . . and if you ever commit such a breach of your orders, I will forgive you also."[27]

This action, and his disobedience of Admiral Parker's signals during the Battle of Copenhagen, when Nelson chose not to break off battle as his less aggressive commander had instructed, made Nelson a household name in England. More important, these actions reinforced in his mind the great value of imparting to his own subordinates the wealth of tactical and operational knowledge he had spent decades learning. The actions also convinced him of the need to empower them with confidence that he trusted them to act on their own initiative when tactical circumstances dictated.

Five months after the Battle of Cape St. Vincent, Admiral Jervis sent Nelson, now a rear admiral, with a detached squadron to attack the town of Santa Cruz de Tenerife in the Canary Islands. To ensure that his plans were well understood, Nelson summoned his captains to the *Theseus* for conferences no fewer than four times: once before sailing and again on 17, 20, and 21 July.[28] They discussed the disposition of forces and assets, Nelson's tactical plans, and the desired goal at the end. But Nelson reassured his captains that

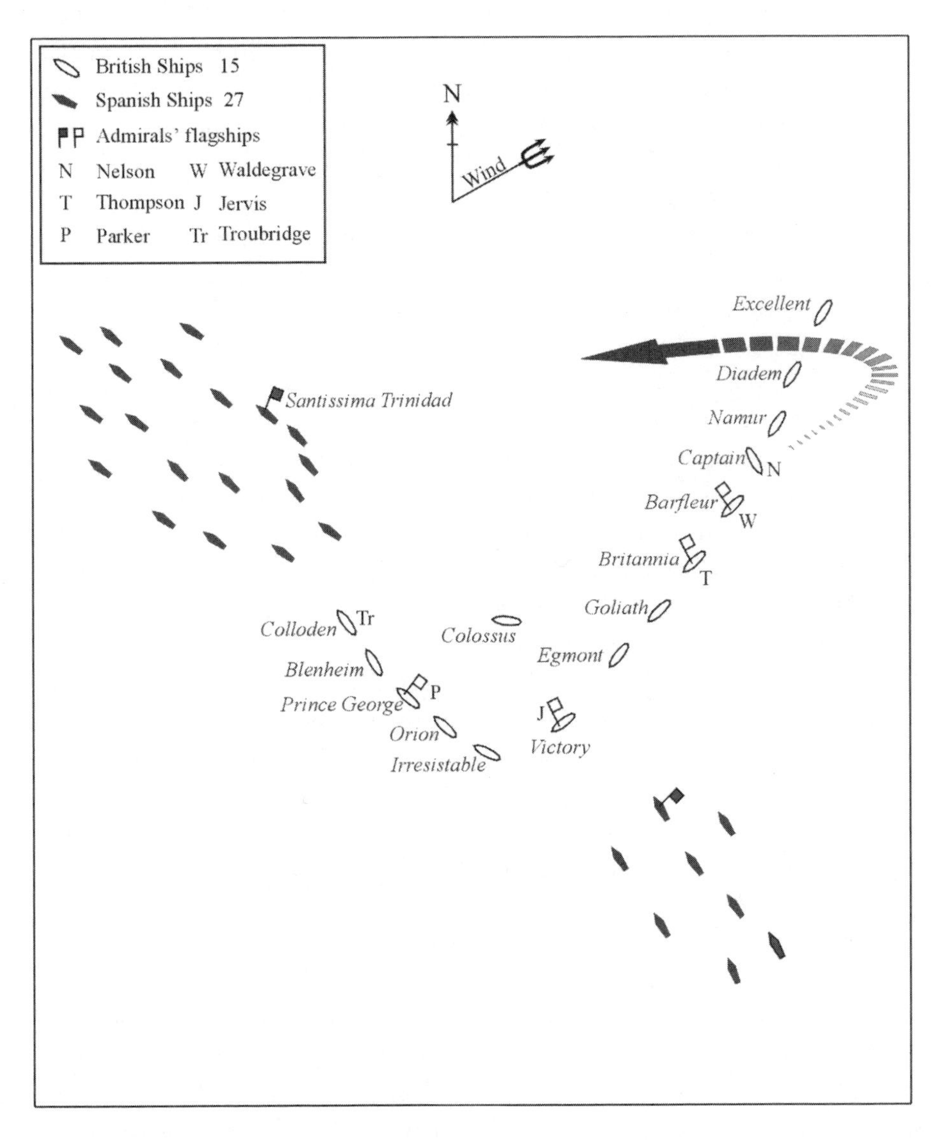

The British Van Attacks at Cape St. Vincent

as circumstances dictated, they were trusted to use their initiative and would be given considerable tactical latitude. To Captain Troubridge, for example, who was to lead the forces landed at Santa Cruz, Nelson advised that he could "pursue such other methods as you judge most proper for speedily effecting my orders, which are to possess myself of all cargoes and treasures which may be landed in the Island of Teneriffe [*sic*]."[29]

The attack on Santa Cruz was a miserable failure—the glaring aberration in Nelson's career—and cost him 343 casualties, either killed or wounded, and the loss of his own right arm. Aside from the inadequate force available, three factors largely outside of Nelson's control caused initial setbacks. These were the state of the sea off Santa Cruz during the summer, unfamiliarity with the mountains behind Santa Cruz, and the prevailing July heat. But the defeat itself stemmed from Nelson's response to those setbacks: allowing his own and his captains' pride to prevail over sound judgment, he made a valiant but ill-considered departure from his usual maneuverist instincts by launching a frontal attack on the enemy's *strongest* point.[30] Although he placed no blame on others, and no one placed it on him, the defeat at Santa Cruz was a bitter pill for Nelson to swallow.

During both the Battle of the Nile and the Battle of Copenhagen, Nelson continued to bring his captains and officers into his confidence. He would brief them fully on his intentions, and convey his trust in their ability to respond to unforeseen circumstances by looking for and making the most of opportunities without waiting for orders. He took communication, training, and trust very seriously, and remarked with all honesty after the Nile that without knowing and trusting his captains, "he would not have hazarded the attack: that there was little room, but he was sure each would find a hole to creep in at."[31] Nelson's statements of total trust in his captains can be contrasted with Wellington's difficulty with delegation[32] and Napoleon's statements in which he expresses utter contempt for his marshals. "These people think they are indispensable," the French emperor once harshly said of his marshals. "They don't know I have a hundred division commanders who can take their place."[33]

Even during the critical Trafalgar campaign of 1805, when the weight of the British people's faith in him caused him acute anxiety, Nelson never displayed Wellingtonian concern at the ability of his commanders, much less treated them with Napoleonic disdain. But then he never shared Napoleon's petty guarding of military knowledge and fear of sharing glory. On the contrary, Nelson wanted his captains to know as much as possible and went to great lengths to pass on his accumulated wisdom. Though he initially knew a minority of the captains in his new command, and realized that not many had ever participated in fleet battle, he immediately took steps to school them in a wide range of tactical options and to familiarize them with the particular plan he intended to execute (these procedures being examples of the so-called "Nelson touch").[34] In a series of conferences disguised as dinner parties on his flagship, he gave tactical lessons that distilled complex maneuvers down to a level that each captain could understand. One lieutenant on the *Victory* later recalled that "the frequent communications he [Nelson] had with his admirals and captains put them in possession of all his plans" and made his intentions known to virtually every officer of the fleet.[35]

Nelson also stressed to them the importance of flexibility. "Something must be left to chance," he told them in a tactical memorandum of 9 October 1805; "nothing is sure in a Sea Fight beyond all others."[36] He added that, if communications became confused during battle, they need not worry unduly because "no Captain can do very wrong if he places his Ship alongside that of an enemy." There, he knew, the better British rate of fire would really count. To Admiral Collingwood, his second in command, he wrote a marvelous note that so perfectly represents the type of command now called Directive Control that it deserves to be read by every maneuverist: "I send you my Plan of Attack," he wrote, "as far as a man dare venture to guess at the very uncertain position the Enemy may be found in. But, my dear friend, it is [designed only] to place you perfectly at ease respecting my intentions, and to give full scope to your judgement for carrying them into effect."[37]

Critics may assert that when Nelson made these expressions of trust, he knew he would be in close proximity to his subordinates during crises and could always intervene if someone did something contrary to his intent. That is true; but certainly Nelson's trust in his officers and his willingness to delegate extended far beyond the tactical level to circumstances where, indeed, Nelson could *not* have exercised spontaneous or even prompt command authority if things went wrong.

In August 1799 Lord Keith left the Mediterranean theater for several months. In his absence Nelson became acting commander in chief of the powerful but widely spread Mediterranean Fleet. Through his deputy, Rear Adm.

John Duckworth, Nelson wasted no time in conveying to his subordinate captains his warmest affirmations of trust. He explained to Duckworth that he would not deluge his deputy with detailed written instructions because he felt the fullest confidence in Duckworth's abilities and judgment. A single line of written direction here and there should, Nelson said, be treated as whole pages from others.[38] Duckworth should rest assured that his "well known judgement and abilities to act as circumstances may require" gave him full operational-level freedom.[39] Instructing his captains to distribute the fleet among Oporto, Lisbon, Cadiz, Minorca, Naples, Sicily, Malta, and the Levant, Nelson allowed them the same marvelous operational latitude. "In giving this command," he said, "I know who I trust to [*sic*], and that it is not necessary to enter into the detail of what is to be done."[40] As Nelson hoped, it boosted morale and increased overall effectiveness. And no captains betrayed his trust.

Nelson's stated intention to keep instructions to a bare minimum is reminiscent of his immediate response to the French fleet's escape from Toulon in April 1805. As soon as he learned that the French had slipped through his blockade, he sent only two signals: one to inform his subordinates that the enemy was at sea, and another for his subordinates to prepare for battle.[41] Commenting on Nelson's minimalist approach to signaling, a marine officer who watched him send these two signals later wrote that it was "his usual way."[42] Nothing more was needed; Nelson's fully trusted and superbly trained captains would know what to do.

Effective exercise of this command style and of reconnaissance pull, for that matter, is largely determined by the commander's willingness to let subordinates depart from orders if tactical or operational circumstances make this absolutely necessary. Slavish adherence to written instructions, if any exist (and commanders like Nelson tried to keep them to a minimum), would prevent subordinates from following their reconnaissance or using their initiative. Fear of punishment for not following orders entirely or exactly as stated would severely inhibit ingenuity and confidence. With this in mind, Nelson believed that all officers, while paying full and proper respect to the superiors they served, had a higher duty to respond to changing circumstances in the manner their superiors would adopt if they were present. As he explained to Earl Spencer in November 1799:

> Much as I approve of strict obedience to orders—even to a court martial to inquire whether the object justified the measure—yet to say that an Officer is never, for any object, to alter his orders, is what I cannot comprehend. The circumstances of this war so often vary that an Officer has almost every moment to consider: what would my superiors direct, did they know what is passing under my nose?[43]

This position may seem extreme, but one should note from Nelson's word "alter"—he did not say "disobey"—that he was certainly not advocating disrespect for authority, insubordination, or needless departures from orders. He made one or two bad choices in his career, but he *never* departed from orders with casual disinterest. He would not have tolerated his captains doing so either. In fact, by stressing both respect for authority *and* the importance of initiative and trust, he was advocating the faithfulness of subordinates to the *intent* of their commanders, who should feel relieved that their juniors were doing their will, one way or another.

Maneuver

Proponents of Maneuver Warfare happily call themselves maneuverists because they believe that skillful and effective maneuvering is the cardinal feature in their warfighting style. By maneuver they mean something more than just the movement of forces. Merely moving reinforcements from rear areas to the front does not constitute maneuver; nor does advancing in linear fashion against an enemy line, à la some of Napoleonic land warfare and much of World War I on the western front. Maneuverists use the term far more specifically to denote the movement of forces to gain either positional or psychological advantage over the enemy with the aim of seeking his defeat, even if not necessarily his destruction.[44] Inexpert head-to-head clashes of forces, especially in linear formations, is not covered by this definition.

For more than a century before the age of Nelson, fleet tactics had been based on the line of battle, which involved fleets or squadrons approaching in long but orderly lines. The combatants engaged each other obliquely or in parallel, with each ship seeking to place itself alongside a similar or weaker opponent. Broadsides dominated this linear form of battle, and victory almost always went to the side that possessed the most and fastest guns (usually the better-trained British). Yet victory was seldom complete and usually consisted of the capture or destruction of a small number of enemy battleships (perhaps two or three), while the majority escaped to fight another day.

Nelson was a passionate student of naval tactics.[45] Deeply impressed by the more significant victories gained in the Battle of the Saintes (1782), the Glorious First of June (1794), the Battle of St. Vincent (1797), and the Battle of Camperdown (also 1797), he could not help but notice that tactical innovations often shocked or confused the enemy and thus brought positive results. He consequently adopted many of his predecessors' ideas as he developed his own theory of warfighting, one that placed great emphasis on a highly effective type of maneuver: concentrating all his focus and strength on a portion of the enemy's line (usually the most vulnerable section), separating it from possible support, defeating it in detail, and then dealing with the rest.

This idea is remarkably similar to the Maneuver Warfare concept of the *Schwerpunkt* (focal point), which refers to the concentration of focus, effort, and strength at a particular point in space and time. Using all available intelligence as well as reconnaissance pull, the commander should determine the most effective place to direct his effort and forces in order to have the maximum effect on the opponent. He should then use skillful maneuver to engage the enemy at that point, ensuring that the strike comes at the most useful time, usually to maximize surprise and preempt enemy reinforcements.[46] Only two of the many famous warriors to emerge in the French revolutionary and Napoleonic periods stand out as consistent practitioners of this idea: Napoleon and Nelson.

Like Napoleon, Nelson clearly aimed for complete victories, and used maneuver and focal points as vital tools. He conducted some bold squadron and fleet maneuvers to gain positional advantage over, and concentration against, carefully selected enemy elements. Once when recommending a daring attack, Nelson told a cautious senior commander, "I am of the opinion that the boldest measures are the safest; and our Country demands a most vigorous exertion of her force, directed with judgement."[47] His definitively vigorous maneuvers included the Battle of St. Vincent, when he pulled his vessel (and several supporting ships) out of the line without instructions to cut off a Spanish movement; the Battle of the Nile, when he sailed down on the

French fleet from the unexpected end of the line and placed ships inside where the French thought the British would never go; and the Battle of Copenhagen, when he sailed along the Danish line from an unexpected direction (up a shallow channel located behind a sand bank, where Danish firepower was weakest). These maneuvers all included great risk, audacity, surprise, and concentration, as did Nelson's greatest fleet maneuver: his bold but very dangerous two-squadron attack to break the combined Franco-Spanish line off Cape Trafalgar. Through skillful maneuver at Trafalgar, he concentrated twenty-seven British ships against twenty of the enemy's and, through British superiority in ship-to-ship tactics and a faster rate of fire, gained his most impressive victory.

Thus, Nelson rejected or at least injected flexibility into the traditional line-ahead approach and maneuvered to gain positional advantage. He believed that his own tactics—which exploited the elements of speed, surprise, and concentration of fire—would greatly enhance the likelihood of attaining the enemy's destruction or collapse because of the *psychological* shock value. In short, he became convinced that his bold attacks would (to quote again from the modern Marine Corps definition of Maneuver Warfare) "shatter the enemy's cohesion . . . and create a turbulent and rapidly deteriorating situation with which the enemy cannot cope."[48]

Collingwood later acknowledged that Nelson's warfighting style simply overwhelmed enemies, who were unable to think of effective responses before it was too late. "It was indeed a charming thing," he observed, recalling Nelson's tactics at the Battle of the Nile.[49] "It was the promptitude, as much as the vigor of the attack, which gave him the superiority so very soon; the Frenchman found himself assailed before he determined how best he should repel the assault." On another occasion Collingwood recalled that "an enemy that commits a false step in his [Nelson's] view is ruined, and it comes on him [the enemy] with an impetuosity that allows him no time to recover."[50]

Nelson's emphasis on audacious and fast moves designed to confuse and hopefully paralyze enemy commanders is a good early example of the modern maneuverist concept of rate of movement, or tempo—"[I] can't bear tame and slow measures," he once told his wife.[51] This attitude proffers that "all actions cause reactions" and the unit or person who *acts* consistently at a faster pace than the opponent will force him to *react* defensively and without time to advance his own plans or create adequate responses. This concept is now referred to in military circles by the buzzword OODA Loop, which stands for the sequential steps in any rational decision: Observe, Orient, Decide, Act.[52] According to U.S. Air Force veteran colonel John Boyd (who first articulated the concept in terms of its role in warfighting), the person who goes through this series of steps (or *loop*) more quickly than his opponent will be "inside the opponent's decision cycle" and thus able to prevent his opponent resisting effectively.[53]

It would be anachronistic to say that Nelson understood combat in these modern, sophisticated terms. He naturally had no idea that he was conducting time-competitive OODA Loops to get inside his enemy's decision cycle. Yet the general idea is the same: he did know he was hitting the enemy with so much speed, strength, and cunning that he was rendering him unable to see through the chaos and disorder in time to plan responses. Thinking he might catch Napoleon's fleet at sea in 1798, Nelson had his squadron constantly ready to fight at a moment's notice, capable of striking so suddenly that the enemy would not have his own ships ready.[54] Likewise, demanding an immediate attack that would take the Danes by surprise and rob them of any initiative, he said before the Battle of Copenhagen: "Let it [the attack approach] be by the sound, by the belt, or anyhow; *only lose not an hour.*"[55] He clearly understood the psychological impact of rapid, audacious, or unusual moves. It inflicted confusion and mental paralysis on the enemy.

In behavior unusual for this period (in which the order and rationality of the Enlightenment was still pervasive), Nelson—like the master of combat psychology, Napoleon—liked the idea of inflicting this chaos: "On occasions we must sometimes have a regular confusion," Nelson said before Copenhagen. Inflicting and exploiting that confusion should be a standard aim in these situations.[56] This viewpoint explains why he had struck so suddenly at the Nile, even though it forced him to fight a night battle. Darkness was the ultimate source of confusion—as he told Lord Howe, "night was to my advantage."[57] The desire for psychological paralysis, even if only temporary, also explains Nelson's comments to a confidant shortly before the Battle of Trafalgar: "I will tell you what I think of it [the maneuver he planned to use]. I think it will surprise and confound the enemy. They won't know what I am about."[58]

Annihilation?

One *might* think that the remarkable similarities between Nelson's warfighting style and the modern Maneuver Warfare paradigm end once the concept of destruction is analyzed. Biographers of Nelson usually insist (to quote one) that he "always sought the annihilation of his enemy," meaning that he aimed for nothing less than the physical destruction of all engaged vessels and, by logical extension, the crews that kept them functioning.[59] This attitude would certainly put him at odds with most modern maneuverists. They argue that one need not destroy the enemy physically if, through applying the principles of Maneuver Warfare, one can so wreak havoc with the enemy's ability to resist that he becomes irrelevant as a threat or as a restriction on national will. In other words, it is not necessary to *destroy* an enemy to *defeat* him.

A cursory survey of Nelson's words and deeds suggests that he did indeed want to destroy his enemies. After the engagement with the *Ça Ira,* he was bitterly disappointed that Hotham had not ordered a pursuit of the French fleet. Fourteen British ships had *captured* only two French, yet Hotham said: "We must be contented. We have done very well."[60] Nelson insisted that had they taken ten ships and allowed the eleventh to escape, he would not have considered the victory total.[61] Similarly, for months after the Battle of the Nile (which established a grim record for its level of destruction and damage inflicted by British warships on an enemy fleet), he regretted that two French ships of the line and two frigates had managed to escape. He took great delight in learning of their eventual capture.

Nelson often used blunt language to describe his intentions and despised partial measures—"half a victory would but half content me," he told one friend on 6 September 1805.[62] Reassuring the same person three weeks later that he was not depressed by his failure to catch the Franco-Spanish combined fleet, Nelson added that "my mind is calm, and I have only to think of destroying our inveterate foe."[63] The word *annihilation* had also begun creeping more frequently into his letters and dispatches after the Peace of Amiens collapsed in May 1803, and he realized he would have to defend Britain in a final showdown with Napoleon's fleet. Arriving off Toulon in July 1803 to watch a French fleet within the port, he stated: "My first object must be to keep the French fleet in check and, if they put to sea, to have force enough with me to *annihilate* them."[64] When it appeared in April 1805 that the enemy fleet might be engaged by a British force under Adm. Alan Lord Gardner, Nelson wished him luck and hoped that he would "annihilate" them.[65] Even during a period of home leave in late August 1805, only seven weeks before Trafalgar, he expressed hope that the combined Franco-Spanish fleet, if engaged by another British admiral in his absence, would be "met with and annihilated."[66] And on 5 October 1805, less than three weeks before the climactic battle, Nelson informed the first lord of the Admiralty, Charles Middleton, first Baron Barham, that he hoped more British ships would soon arrive to spoil the plans of the Franco-Spanish force and that "as an Enemy's Fleet they may be annihilated."[67] This was real fighting talk. Still, what did Nelson mean by annihilation?

The word *annihilation* is almost identical in meaning and usage to the German word *Vernichtung* (now often translated as "extermination," a far more specific and cruel meaning than its earlier rendering as "destruction") used by the Prussian war philosopher Clausewitz to describe the type of battle that was supposedly best able to deliver total victory. During the Napoleonic period, both *annihilation* and *Vernichtung* usually denoted reduction of something to the state of nothing; that is, an object's physical destruction or disintegration. An action that might fit this meaning, for example, is the grinding of a dense substance into a fine powder using a mortar and pestle.

Yet one should note that the word *annihilation* was used with other, less absolute meanings. In 1799 Nelson felt distressed at a letter he received from St. Vincent, in which the old admiral harshly criticized Josiah Nisbet, Nelson's adopted son (we would today merely say stepson). He was ignorant, lazy, and inattentive, St. Vincent complained, "and had he not been your son-in-law [*sic*] he must have been annihilated months ago."[68] Clearly the admiral wasn't trying to say that Josiah would have been physically destroyed (that is, killed or murdered) if he were not Nelson's adopted child, only that his naval career would have been promptly terminated. Likewise, in August 1799 Nelson anticipated that a British fleet under Lord Keith would "annihilate" the massive Franco-Spanish fleet sailing somewhere in the Mediterranean.[69] He knew full well that the best Keith could hope for was to hurt the enemy fleet and hopefully prevent it landing an invasion force on Minorca. Yet Nelson used the word *annihilation* for a possible battle to be fought by Keith —someone he considered cautious and conventional and defensive minded—

that would undoubtedly not equal, say, the Battle of St. Vincent two years earlier. And even that great clash had never come close to achieving the total destruction of the enemy.

Many who study war believe that both Clausewitz and Nelson advocated inflicting the maximum bloodshed on the battlefield; that they measured success in terms of a friend-foe casualty calculation, and considered victory was won by the side that inflicted unsustainable or unbearable losses on the enemy. This interpretation overstates Clausewitz's argument. He did write such bloodthirsty-sounding things as "Destruction of the enemy's military forces is in reality the object of all combat" and "What is overcoming the enemy? Always the destruction ["Vernichtung"] of his military force, whether it be by death, or wounds, or any means."[70]

Yet a careful reading of Clausewitz shows that he never saw bloodletting as the end in itself, merely as a means to the end, and he never said that it was the *only,* or even most desirable means. The objective Clausewitz advocated was the imposition of one's will over the enemy so that he would cease to threaten safety and interfere with the attainment or continuation of national policies. This could be achieved through a variety of nonmilitary means, particularly diplomacy; but if battle had to be joined, the desired goal was still not destruction for its own sake or as a measure of victory, but the eradication of the enemy's ability to spread fear, cause harm, or inflict loss of life. The desired goal of battle was "disarming the enemy" and rendering him nonfunctional. "Destruction," Clausewitz explained, means "reducing a military force to such a state that it is no longer able to prosecute warfare."[71]

Lord Nelson's position, it would seem, has been likewise overstated by at least a small degree. He never set down his military ideas in a thorough and systematic fashion. Also, like both Wellington and Napoleon, he certainly never wrote anything comparable to Clausewitz's masterpiece, *On War.* Scholars thus have had little alternative but to pick out passages scattered throughout Nelson's letters and dispatches and to compare them to the conduct and outcome of his battles. When scholars do so, they notice that Nelson sometimes said certain things that eventually occurred. They logically deduce, therefore, that the events happened in their precise manner because Nelson planned them to do so. Thus, if Nelson used certain words to describe a forthcoming battle, and that description matched the actual event, he must not only have chosen his words carefully and with precise meanings in mind, but also resolved to put them into action. According to this logic, Nelson must have really aimed to destroy physically all enemies in each of his major engagements.

It is clear that Nelson aimed for total victory and in battle showed little concern for resulting enemy deaths. "I hope and believe that some hundreds of French are gone to hell," he told Emma Hamilton after a raid on Boulogne in August 1801.[72] But that does *not* mean he desired the enemy's total physical annihilation, despite Nelson's frequent rhetoric to this effect after the collapse of the Peace of Amiens in May 1803. He was in fact a remarkably complex mixture of ruthlessness and humanity, a man who mercilessly unleashed the full power of his forces at the start of battles but, once victory had been attained, usually ended them with an almost paradoxical humanity. Available sources actually indicate that while he did hope his last big battle with the French would end in a victory surpassing his previous achievements,[73] his thinking was similar to Clausewitz's (and a bit more in keeping with modern maneuverists than previously thought): he saw battle as a means of gaining personal glory (which he always craved), but much more important, he recognized that the primary function of battle was the removal of all threats to British safety and interests caused by the enemy's forces. This conclusion need not be achieved by total physical destruction and great bloodshed if the enemy's cohesion, morale, and will to fight could be otherwise shattered and his forces reduced to irrelevance.

Ships of the line were, in any event, remarkably tough vessels, and their complete physical destruction through burning or sinking was not easily achieved. Damaging them to the point of combat uselessness or capturing them for prize money and later addition to the British fleet was easier and usually attempted as one or another battle objective. Throughout the entire Napoleonic period, these actions were also accomplished at a greater rate.

The result of all these actions—capturing, seriously crippling, forcing aground, burning, and sinking—was the removal of enemy ships from any future use in threats, coercion, or combat. That was precisely what Nelson wanted. He saw no real difference in the relative value of any one of these actions (he certainly never expressed greater pleasure in sinking a ship and killing its crew than in capturing them) and considered all these actions "destruction" if the results were ship losses to the enemy.

To illustrate Nelson's position: throughout late 1804 he complained bitterly about the many pirates and privateers that interdicted British shipping and caused other problems to his fleet. He informed one of his captains that he wanted the pirates "kept in pretty good check, and destroyed the moment they attempt to proceed without the protection of neutrality."[74] Yet he mentioned to another officer his hope that he would soon "*capture* or destroy more of those piratical Privateers."[75] Likewise, when Spain formally sided with France against Britain in November 1804, Nelson instructed his captains and commanders to use their "utmost endeavour to *capture, seize,* burn, sink or destroy" any Spanish vessels.[76]

This caveat also applied to fleet actions. Capturing or sinking ships was all the same to Nelson, so long as they were robbed from enemy service and consequently eradicated as a threat and nuisance. On 23 July 1805, for example, he notified Lord Barham that Nelson's ships had almost caught the enemy fleet in the West Indies and that, even though the outcome of battle "would have been in the hands of Providence," he humbly believed that "the Enemy would have been fit for no active service after such a Battle."[77]

This view of "destruction"—that it includes the capture and crippling as well as the sinking or wrecking of enemy ships—allows us to make more sense of Nelson's aforementioned complaints about Admiral Hotham's inconclusive engagements in 1795. Clearly Nelson's desire to have "taken" ten out of eleven ships does not refer specifically to their physical destruction, but to their removal from French service through capture, permanent damage, or sinking. It is the same with Trafalgar. On the day of battle, Nelson asked Captain Blackwood what he would consider a victory, to which the frigate captain replied: "If fourteen Ships were *captured,* it would be a glorious result."[78] Nelson replied that he would not be satisfied with anything short of twenty. Again, this demonstrates that in the great commander's mind, the important thing was to rob the enemy of effective naval strength, and this could be accomplished by physical destruction *or* capture.

In fact, if one looks at the Trafalgar campaign of September and October 1805, one does not find Nelson aiming to destroy the same percentage of ships as he had during the Battle of the Nile in August 1798. Then, he had captured and destroyed eleven of thirteen ships of the line and most frigates and smaller craft (thereby reducing Napoleon's Egyptian fleet by almost 90 percent).

In 1805 Nelson often threatened destruction and annihilation and wanted a battle of unsurpassed totality, and never shied away from inflicting a lot of bloodshed. Yet, he realistically aimed for the level of accomplishment that would secure mastery of the seas for Britain, reassure the British people that their islands were safe,[79] remove the threat of invasion or mischief by Napoleon, and provide security to Britain's allies. He measured this level of accomplishment at twenty enemy ships of the line robbed from Napoleon's fleet of thirty-three.[80] This number means that he aimed to reduce (or "degrade," to use today's jargon) the combined fleet by about 60 percent, a remarkably ambitious undertaking (at Trafalgar he was slightly outnumbered, and none of his illustrious predecessors had taken twenty ships in all their battles put together). Yet it simply does not support the claim, often made by writers, that "the total destruction of the enemy force" had become the admiral's goal.[81]

As it happened, the British fleet at Trafalgar lost none of its own ships but took or destroyed a total of nineteen French and Spanish ships. Had he lived, Nelson would have been ecstatic about the safety of his fleet and the destruction of the enemy. "I would willingly have half of mine burned to affect their destruction," he had claimed during a moment of gloom nine months earlier, in keeping with a statement he had made in 1799 that "victories cannot be obtained without blood."[82]

This Franco-Spanish destruction rate of 57 percent may sound highly attritional and make some maneuverists squeamish. Yet one must remember that this percentage refers to ship, *not human,* losses. For only 1,690 casualties of their own, the British killed and wounded 6,953 Frenchmen and Spaniards and took twice that number prisoner.[83] In terms of the numbers present at the battle (over fifty thousand) and the scale and significance of the victory gained, these are certainly not horrific losses. They compare favorably—almost identically—with the battlefield casualty rate (8 percent friendly, 20 percent enemy) achieved by the German Wehrmacht during their successful Blitzkrieg campaigns from 1939 to 1941.[84] (For these campaigns the Wehrmacht has been hailed by modern maneuverists as fine exponents of their warfighting style.) And the Trafalgar casualty rates compare extremely well with those of many Napoleonic land battles, which were more attritional than many people think. At Albuera, for example, the British and French both suffered 44 percent casualties, and at Waterloo, 24 and 35 percent respectively.[85] In short, although Nelson clearly realized that taking ships inevitably meant killing many crewmen, he was no attritionist.

Conclusions

Nelson, even though he was fighting at sea and not on the land, developed a warfighting style very similar to that now called Maneuver Warfare and hailed by its many advocates as the most intelligent and effective means of waging war at the tactical and operational levels. Nelson saw warfare in the same uncomplicated terms as George Patton. "War is a very simple thing," the best Allied maneuverist of World War II said, "and the determining characteristics are self-confidence, speed and audacity."[86] Were Nelson listening from the grave, he doubtless would have replied: "Hear, Hear!"

Nelson learned by studying battles fought in previous generations by the naval heroes he idolized and by experimenting with feelings and ideas arising from instinct and intuition. His success, he knew, depended on psychological as well as physical attack, and he endeavored to master both. Endowed with an aggressive spirit and an awareness that partial victories only meant more battles to fight at later stages, he aimed for complete victory. That does not mean he aimed for total physical destruction, despite his own bombast and the later claims of many biographers. He aimed for what would shatter the enemy's will to fight, give Britain and its allies safety and freedom to carry out their national policies, and deny the enemy the ability to cause further mischief. That meant totally defeating the enemy, not necessarily totally destroying him. Yet, Nelson—a complex mix of ruthlessness and humanity—never flinched from hitting the enemy cunningly and ruthlessly to inflict as much damage and bloodshed as possible for the minimum loss of his own seamen's lives. He instinctively used virtually all the concepts now extolled by maneuverists—reconnaissance pull, the application of strength against weakness, Directive Control, focal points of effort, maneuver, tempo, and the aim of achieving victory through preempting the enemy's decisions and shattering his will to fight—to gain the stunning victories that gave his country naval supremacy for a full century. He never knew to explain his actions in these terms, but a maneuverist Nelson surely was.

5

Nelson and War on Land

LORD NELSON S talents as a naval warrior have not been exceeded or exaggerated. He was extremely good and rivals Napoleon for the honor of being their era's greatest champion. Napoleon mastered warfare on land, winning many battles and campaigns with an artistry that the military theorist Clausewitz would later describe as "great genius." Nelson, meanwhile, mastered warfare at sea, winning his own battles and campaigns with so much instinctive flair and totality that the naval theorist Mahan would describe him as "the embodiment" of sea power.

Yet the environments in which land and sea battles take place are very different. To use modern professional jargon, the *battle space dynamics* of land and sea are so distinct that planners and practitioners cannot utilize all the same strategies, operational art, and tactics when pursuing victory in each environment. Terrain is a critical factor in land warfare. Mountains, swamps, deserts, and rivers form obvious obstacles to armies and their weapons and supplies. Even less apparent geographical features—such as a series of hedgerows, a gently sloping hill, or a line of sand dunes—can dramatically influence the way a battle is fought. The sea, on the other hand, seems flat, constant, and only slightly varying in its surface contours, except perhaps during storms or other extremes of weather. Yet the sea is a perilous and an equally awkward battle space. It has currents and tides and subsurface features such as sandbars, shoals, and rocks that significantly impact on the way naval battles are fought. Weather and other meteorological conditions have different effects on each environment. Strong winds will be a nuisance to an army; but in the age of sail, adverse winds often paralyzed shipping for weeks and sometimes proved the decisive factor in battles at sea.

Warfighting tactics and principles also vary. *Turning an enemy's flank* is a basic and often effective move on land. At sea, naval forces really have no comparable move; there are no flanks. *Envelopment* also means something different to a general, who might hope to enclose his enemy and sever his lines of communication, than it does to an admiral, who might want to *double* enemy vessels by firing upon them from both sides. Even the concept of lines of communications differs between the land and sea environments. On land it usually denotes the logistical routes that lead from rear areas to combatants at the front and upon which information, instructions, food, fuel, munitions, and reinforcements travel. At sea there is, once again, no directly comparable concept. Vessels, squadrons, and fleets operate in and along sea lanes that extend from home or friendly ports; but there is no continuous flow of traffic between ports and warships at sea that can be severed in similar fashion to taking or destroying a bridge, road, or railway.

It does not follow, therefore, that someone who masters warfare in one environment will necessarily prove successful in the other. It is well known and frequently reported that Napoleon knew little about sea power and consequently squandered the use of his quite good fleets. The emperor (not the world's most self-critical dictator) naturally blamed his admirals for his lack of success at sea. Lamenting in 1816 that he had been unable to find his own Nelson—whom he greatly admired—Napoleon complained that every time he proposed a naval or amphibious operation his admirals would insist on its impossibility.[1] Their "specialized, technical mentality" and their obsession with currents, winds, tides, and so forth always ruined his plans. "How can a man argue with people who speak a different language?" the great soldier fumed.

This chapter explores Nelson's understanding and conduct of land warfare, which he undertook on several occasions, some of them for extended periods. It seeks to answer several important but hitherto ignored questions: Did he, serving in the Royal Navy, which most people of his time considered the "senior service," understand the basic principles of military—that is, army—tactics? Did he understand the importance of terrain and the other influential features of battlefields? Did he and his subordinate naval forces work harmoniously and effectively with soldiers and army officers? Did he understand and respect the army's role in Britain's overall strategies? To what degree did he contribute to implementing the strategies? Lastly and most importantly, were Nelson's naval talents matched by similar military instincts and abilities, or was he a fish out of water in terms of land warfare?

Early Experiences

Joint operations involve the closely synchronized employment of two or more service branches—for example, navy and army or army and air force—under a single commander with legally vested authority over all force elements in the theater. Military theorists unanimously argue that joint operations prove more effective in most circumstances of modern warfare than operations involving only one service or two or more services but without systematic coordination or unity of command. Jointness is so much a feature of modern warfare that a young cadet from any armed service can rightly anticipate that sooner or later, he or she will be training, working, or fighting alongside members of other services.

Yet this situation is very new, postdating the Second World War. In previous conflicts, including the catastrophe of 1914–18, armies and navies generally conducted separate operations based on different and poorly coordinated strategies. Armies fought other armies, in Britain's case usually after being transported to the Continent or elsewhere with naval assistance. Navies fought other navies, undertaking blockades, coastal protection, commerce raiding, and sea battles but rarely with any concern for what armies were doing elsewhere (except when logistical provision proved necessary). Generals and admirals seldom met and almost never coordinated fully joint operations, let alone as equals. Soldiers and seamen (except for marines—the sea soldiers) never trained together and hardly ever fought together.

Nelson was thus not unique in having very little contact with soldiers and their tactics during the formative years of his naval career. Indeed, throughout the eight years between joining the *Raisonnable* as a twelve-year-old midshipman in March 1771 and his appointment as a twenty-year-old captain of the twenty-eight-gun frigate *Hinchinbrook* in June 1779, his naval experiences were exclusively limited to routine service, including supply and escort duties and a few minor engagements. His other experiences at sea during these years included a brief but valuable spell aboard a merchant ship in the West Indies and participation in an unsuccessful Arctic exploration voyage.

During those years Nelson became a skillful seaman, as his own biographical sketch of many years later attests. He admitted that his time aboard the merchant ship helped make him (at fourteen years old) a "practical seaman."[2] Likewise, his service thereafter aboard a seventy-four-gun ship of the line lying at the Nore (a sandbank in the Thames), along with time spent aboard the seventy-four's supporting cutter and decked longboat of which he gained command, made him a good pilot for those types of vessels. Navigating his longboat through the Medway River and the Swin channel, leading to Chatham from Sheerness, and through the sandbanks of the Thames estuary made him "confident . . . amongst rocks and sands." Similar experiences and growing navigation and command responsibilities during the next few years increased the young man's abilities at sea. By the age of eighteen, he had already mastered a wide range of tasks, soaked up a virtual encyclopedia of advice on naval tactics and war from his mentors, and successfully boarded several privateers and small enemy ships. Yet he had heard very few shots fired in anger and at no time experienced land warfare firsthand.

That seemed likely to change in August 1779. France had declared war on Britain eighteen months earlier, siding with the American colonies in their desire for freedom from King George III. The Caribbean became a theater of war between France and Britain, both of whom had a fairly equal split of possessions there, but with France enviously eyeing Britain's sugar-producing islands. In August 1779 French admiral the comte d'Estaing was reported to have arrived in the Caribbean with a massive fleet, including thirty battleships and one hundred troop transports. The ability of the French to land twenty-five thousand troops sent British colonies in the region into a tailspin. Nelson, then on shore duties in Jamaica, told his old mentor Locker that the island was "turned upside down" by the news.[3] An attack on Jamaica was expected, Nelson explained years later, and

> in this critical state, I was . . . entrusted with the command of the batteries at Port Royal; and I need not say, as the defense of this place was the key to the port of the whole Naval force, the town of Kingston, and Spanish Town, it was the most important post in the whole island.[4]

Nelson exaggerated his role. Although the twenty-year-old officer doubtless had a valuable position to defend, and would have fought with the tenacity and courage for which he would later become famous, the vast majority of cannons and troops were elsewhere and under other commanders far senior to him. As it happened, Comte d'Estaing decided against attempting an invasion, instead sailing north to aid George Washington's army in its fight for freedom. Nelson, who had yet to prove himself in any real combat, thus missed out on what would have been his first experience of amphibious assault or land warfare.

Such a battle's outcome could be predicted; after all, how could seven thousand defenders, including a hastily raised militia, withstand the onslaught of twenty-five thousand invaders? Nelson himself anticipated that his imminent captivity might give him the necessity to learn French.[5] Yet no matter how fierce the fighting might have been if d'Estaing had attacked, and Nelson probably sighed with relief when it became clear the French admiral was not going to, the battle would actually have required of Nelson few new tasks and exposed him to nothing he had not been trained to do. He was in command of fixed positions, with very limited opportunities for tactical maneuver and with none for using his own combat initiative, firing cannons at slow-moving French ships or at troops in boats attempting to land.

Fresh challenges and the potential for new experiences of war presented themselves in early 1780, when Nelson, now twenty-one and captain of the *Hinchinbrook,* received instructions to transport soldiers and stores from Jamaica and land them near the mouth of the San Juan River in Nicaragua. The governor of Jamaica, Maj. Gen. John Dalling, and the secretary of state for the American colonies, Lord George Germain, had responded to Spain's hostile entry in the war with the grandiose plan of severing and hopefully supplanting the Spanish colonies in Central America. This would be accomplished by sending an army expedition up the San Juan River and across Lake Nicaragua to the city of Granada, Nicaragua's largest town, itself within striking distance of the Pacific coast. A navigable waterway might thus be opened between the Atlantic and Pacific Oceans, which would sever Spanish communications between its colonies to the north and south while giving Britain new and strategically useful holdings.

The plan involved Capt. John Polson (promoted to lieutenant colonel for the mission) of the Sixtieth Foot, the King's American Regiment,[6] leading an advance in boats and canoes up the San Juan River and then an assault upon Fort San Juan (El Castillo de la Immaculada Concepción), which guarded the approach to Granada. Polson would have his own troops and a newly raised regiment, the Seventy-ninth Foot, assisted by released convicts, freed slaves, and a motley collection of Jamaican vagrants. Polson would also gain the valuable assistance of an efficient Irish officer from the Fiftieth Foot, Capt. Edward Despard, who would assume engineering and artillery responsibilities. Young Nelson (whom Polson initially considered "a light-haired boy" of no consequence[7]) also joined the team, assigned to escort the convoy of troopships and smaller vessels to the mouth of the San Juan River.

Nelson, ever pleased to be kept busy, happily received his instructions even though his own role promised few opportunities for action and glory. He was supposed to have no further participation after landing the troops and stores. He hoped that Polson's expedition would succeed, but felt some doubt. "How it will turn out," he wrote to Locker, "God knows."[8] He would soon find out, to his grave disappointment.

He transported the expedition to the aptly named malarial Mosquito Shore, an eventful six-week voyage full of bungled planning (not by him). He then decided, contrary to his instructions, that he would not return to Jamaica. After watching several of Polson's boats capsize at the mouth of the San Juan in a scene of embarrassing confusion and incompetence, Nelson realized that the expedition desperately needed a skilled boatman. He consequently offered, on his own initiative and exceeding his orders, to take charge of the boats and lead them up the river.[9] He also offered to contribute thirty-four seamen and thirteen marines to the five-hundred-man force. Polson was delighted and gladly entrusted Nelson with leading the expedition up the horrid river. "I want words to express the obligations I owe that gentleman," Polson later wrote, firmly convinced that the "light-haired boy" was no ordinary seaman.[10]

Nelson's first taste of land warfare was anything but enjoyable. The stifling humid days proved exhausting, with the overcrowded boats being paddled against the current, hauled over shoals and sandbanks (on which they often grounded), and loaded and unloaded each night as camps were established. The nights were hardly restful; mosquitoes and other insects tormented the weary and constantly wet travelers. They found the frog croaks, bird screeches, monkey chatter, and other noises of the ever-thickening rain forest both scary and disturbing to their attempts at sleep. Worse still, the inadequate supply of provisions prompted the Indian cooks to begin supplementing rations with fish, lizards, large rodents, and even monkeys.

On 9 April 1780, after thus struggling upstream for two weeks and seventy miles, the forward scouts discovered a Spanish battery on the small island of Bartola. It obviously served as an outpost to Fort San Juan, still several miles upstream, and could not be bypassed. It had to be taken. Nelson, whose energy, selflessness, and sound ideas had earned Polson's admiration, took part in planning the attack.

No interservice rivalry existed; this was a united team working toward one objective, and Nelson, as keen to learn as he was to help, found himself constantly included in all decision making.

For this attack two small groups would storm the battery simultaneously at first light, thus exploiting the element of surprise. The battery was so small and low that no escalade (an attack up-and-over walls using scaling ladders, grapnels, and other such equipment) would be needed. The battery could be attacked directly. Nelson, accompanied by Despard, would already have two boats upstream of the battery, so they could sweep down upon it with the current, while another group provided covering fire with muskets and light cannon from the bank opposite the battery. A weak third group would penetrate the jungle and work its way behind the battery to prevent the defenders escaping and warning the main fort further upstream.

Surprise works wonders in battle *if* it can be gained, which is usually far more difficult than planners think. It proved so on this occasion. At dawn's first light, a shouted challenge in Spanish told the attackers—who had failed to get their boats as far upstream as they hoped—they had been seen. With surprise gone, Nelson's seamen rowed frenetically toward the island, with Nelson drawing his sword and scrambling into the bow to be the first ashore. Jumping down he sank deeply into mud and had to struggle, with musket balls flying around him, before he could pull free of his stuck shoes and dash forward, in bare feet, at the enemy.

Nelson had doubtless been offered a musket, but preferred his trusty naval sword (as it happened, he always remained a hopeless shot with a firearm) and perhaps a pistol.[11] Many of his seamen, with the probable exception of his marines, also fought with the regular hand weapons of their service—axes, pikes, and cutlasses—even though this choice put them at a disadvantage against their firearm-wielding foes. Yet Nelson's seamen fought fiercely side by side with Polson's musket-armed soldiers, inflicting equal terror and damage. They certainly demonstrated the same craving for victory.

The terrified Spanish quickly surrendered, although not before sending a warning to the fort upstream. The intense but small-scale battle reminded Nelson so much of the close quarters ship-to-ship action—with a frontal assault, a short burst of frenzied violence by men tightly crowded together, and an enemy blazing away from almost point-blank range—that he later proudly wrote that he had successfully "boarded" the enemy position.[12]

Nelson's next battle, against the Spanish fort itself, in no way resembled a naval fight. Excited by success against the Spanish outpost, the young sea officer urged a rapid advance upstream to the fort and an immediate attack on its garrison.[13] Even after soon seeing the fort on a steeply sloping hill, with strong stone walls and ramparts supporting many cannons, he still insisted that it should be stormed forthwith in much the same way as the outpost had been.

This was the way seamen boarded enemy vessels at sea: up close and personal, attacking audaciously and with frenzied violence. The aim of boarding was to get past the enemy guns and to inflict an unsustainable level of bloodshed on the defending crew. Any responsible captain would surrender rather than lose too many lives.

But this sort of approach was not a wise way for soldiers on land to fight, particularly when the enemy enjoyed relative security behind strong unbreached fortifications—with guns that had interlocking fields of fire, built on ground difficult for attackers from almost all directions. A frontal assault under such circumstances, no matter how courageous and tenacious, was unlikely to prove successful—at least without horrendous losses. Fort San Juan, El Castillo de la Immaculada Concepción, was one ship that Nelson and his fellows should not attempt to board.

Polson therefore reined in his young subordinate, telling him that they could only attack by escalade or by breaching the walls with cannons. In the absence of escalading equipment, their only option was to lay siege to the fort with artillery pieces after situating them on commanding heights nearby.[14] The fort could not hold out indefinitely because it had no enclosed source of water. Any water inside the fort would become rancid within a few days of stifling heat. The defenders would soon capitulate, Polson said, especially after the expected British reinforcements arrived with fresh provisions and additional ammunition for their small 4-pounder guns.

Many Nelson biographers, enthralled by their hero's courage, have criticized Polson's tactics and insisted—entirely incorrectly—that the immediate frontal assault recommended by the young seaman would have overwhelmed the Spanish defenders and brought victory. It would supposedly also have prevented the loss of many lives caused by tropical illnesses made worse by the heavy rains that soon began to fall.

Clarke and M'Arthur, authors of one of the earliest biographies of Nelson, claim that Polson's decision to ignore Nelson's advice was caused by "misunderstandings, oppositions and delays," which were the "ruin" of many military operations.[15] One recent biographer wrote this incredible sentence: "Napoleon would undoubtedly have acted as Nelson would have done, but lesser minds [presumably Polson's] must always abide by the book of rules."[16] Nelson perceived "the folly of the method," but obeyed Polson's instructions.

That claim is pure nonsense; Napoleon was as instinctive and audacious as Nelson, but he was vastly more experienced in land warfare. Napoleon could have guessed that the attackers' greatest weapon, surprise, had almost certainly been lost (indeed, the fort had been forewarned after the fall of the outpost downstream). He would also have observed, like Polson, that the fort was strong and well formed and sat on a steep conical hill cleared of trees, which made an ideal killing ground with no possibility for the attackers' concealment, camouflage, entrenchment, or sapping. The twenty Spanish cannons and twelve swivel guns, many mounted on bastions with interlocking sweeps, would prevent a river approach (difficult anyway because of rapids alongside the fort) and inflict heavy losses on any force attempting to storm the hill. Those who survived the hail of grapeshot would be picked off by muskets aimed from the safety of the walls and ramparts. And lacking escalading equipment, anyone who reached the fort's unbreached walls would have been left helpless.

In any event, of the five hundred or so men available to attempt such a mad attack, only two hundred were trained regulars, the rest being of limited (and some of no) combat value. The growing sick rate caused by exhaustion and tropical illnesses reduced combat capabilities further. The attackers were thus numerically too weak for a Nelsonian-style assault.

Polson's tactical solution was undoubtedly correct, and a siege was duly established. Getting the British guns into position proved difficult and exhausting, but, under Nelson's and Despard's guidance, at least the gunfire proved accurate and effective. One might think that Nelson, experienced at firing ships' guns only, would have made a poor artillerist on land. Seamen usually did. Used to firing broadsides at close range and at great speed, without excessive concern for precision, they generally found long-range, sustained, and accurate firing hard to achieve. Yet Nelson made an excellent artillerist, prompting Polson to record in his official dispatch that "there was scarcely a gun [fired] but was pointed by him or Lieutenant Despard."[17] Nelson himself remained ever proud of his well-situated and operated batteries, later claiming them to be "a principal cause of our success."[18]

Unfortunately, Nelson did not remain in-theater long enough to see the fort's capture (which took eleven days of siege and occurred only after reinforcements arrived with more shot and shells, the original lot having been expended quickly). Nelson, ordered by Admiral Parker to return anyway, became dreadfully ill after drinking rancid water and had to be evacuated downstream to the coast and from there to Jamaica. He was lucky to leave; many of those who remained succumbed to jungle illnesses and soon perished. Close to death himself, which he believed imminent, Nelson barely survived the journey. Thankfully, careful nursing in Jamaica eventually brought him back to better health.

So what conclusions can be reached about Nelson's first foray into land warfare? He proved invaluable to the expedition, but primarily because of his boat-handling experience. In a spirit of teamwork and wholehearted cooperation, he safely took the army expedition upstream, navigating treacherous waters, rapids, sandbanks, and other hazards with great skill. This is not to demean his contribution to combat. His cannon placement and firing were superb. Moreover, Dr. Benjamin Moseley, the expedition's physician, recalled that Nelson was the first ashore during the assault on the outpost and that his courage in the face of heavy fire inspired the British seamen and soldiers and caused the Spanish defenders to panic and flee.[19] This was the Nelson style, characterized by courage, audacity, and aggression. Having said that, the Royal Navy owes Polson a huge debt of gratitude for ignoring Nelson's advice regarding the storming of Fort San Juan and thus saving the sea officer for his famous battles in later years. In terms of the dynamics and tactics applicable to land warfare, Nelson's advice was uninformed, dangerous, and quite wrong.

Corsica

Nelson's next and most significant participation in land warfare occurred over thirteen years later, shortly after he entered the vast Mediterranean Sea for the first time in the summer of 1793. As already noted, Admiral Lord Hood accepted responsibility for the French port of Toulon after its frightened citizens appealed for protection from the advancing republican army that had already acted brutally in Marseilles. Hood dispatched Captain Nelson in the *Agamemnon* to raise troops in Naples and elsewhere and assist in their conveyance to Toulon to aid in its defense. All their efforts proved inadequate after the young, then-unknown Napoleon Bonaparte expertly trained his cannons on Hood's ships in Toulon and forced them hastily to evacuate the harbor, leaving many French loyalists in the hands of vengeful republicans.

The loss of Toulon and the massacre of loyalists left Hood, Nelson, and their naval fellows gravely upset and with a significant problem. They now had no base from which to conduct further operations against French republican forces in the Riviera and to blockade their fleet in Toulon—only part of which the British managed to destroy as they fled the harbor.[20] With Gibraltar far too distant, logistical demands growing by the week, and shelter from gales and storms needed, only one port and anchorage seemed to meet Lord Hood's requirements: San Fiorenzo in northern Corsica, only a hundred miles from the southern French coast.

This objective matched Hood's ambition of helping Corsican nationalists, under the famous patriot Pasquale Paoli, to liberate their island from French rule. French coastal forts in Corsica were already being blockaded by one of Hood's subordinates, Commodore Linzee.[21] On 4 January 1794, Paoli renewed earlier appeals to Hood for help, offering to place the island under British protection.[22] On behalf of George III, Hood accepted, and duly began planning a joint army-navy campaign to wrestle the island from its French garrisons.

Good relations between the army and the navy would be essential to the campaign. The army had insufficient supplies and stores, and because of the Toulon debacle, it had almost no artillery. It therefore needed not only the navy's logistical support, but also its ships' guns and the crews to haul and man them. The navy, in turn, lacked a sufficient number of marines and seamen to conduct amphibious operations by itself. It needed the army if it wanted to undertake any combat ashore.

Clearing Corsica of French troops proved difficult, costly, and time-consuming, although the first stage, securing San Fiorenzo, took only two weeks. The port's defenses included a strong fort on Mortella Point and a series of strong fortifications at Fornali, further to the north. On 7 February 1794, troop transports escorted by Linzee's squadron disembarked 1,400 British infantry and 120 seamen near the fort at Mortella, under the command of Maj. Gen. Sir David Dundas.[23] On the next day, two of Linzee's warships attempted to bombard the fort but took a mauling from the fort's guns, which fired red-hot shot and killed and wounded sixty seamen.[24] The ships beat a hasty retreat out to sea. Dundas had better fortune. With the help of skilled seamen, who hauled guns with great efficiency, he raised a battery of four guns on the shore. After two days of bombardment, he gained the fort's surrender.[25]

Another force of seven hundred soldiers and seamen, under the command of Lt. Col. (later the famous Lt. Gen. Sir) John Moore, meanwhile besieged the fortifications at Fornali. This siege was only accomplished after seamen dragged light and heavy ships' guns over very difficult terrain from Linzee's landing point.[26] Dundas marveled at their abilities. Indeed, cooperation between soldiers and seamen at the tactical level was excellent.[27] Later, Moore generously recalled how the assistance provided by Capt. Edward Cooke and his crew, particularly the hauling, situating, and firing of guns, contributed significantly to his successes.[28] After two days of their bombardment, Dundas ordered Moore to storm the fortifications.[29] An aggressive bayonet charge helped Moore's troops overcome the strong French garrison of 550 men, of whom they killed or wounded 170 compared to their own casualty total of approximately one-third that number.[30] Recognizing their cause as lost, the French garrison in San Fiorenzo withdrew to the strongly fortified city of Bastia (capital of northern Corsica, situated on the east coast, sixteen miles by road from San Fiorenzo), thus leaving the entire Gulf of Fiorenzo in British hands. Hood had his safe anchorage.

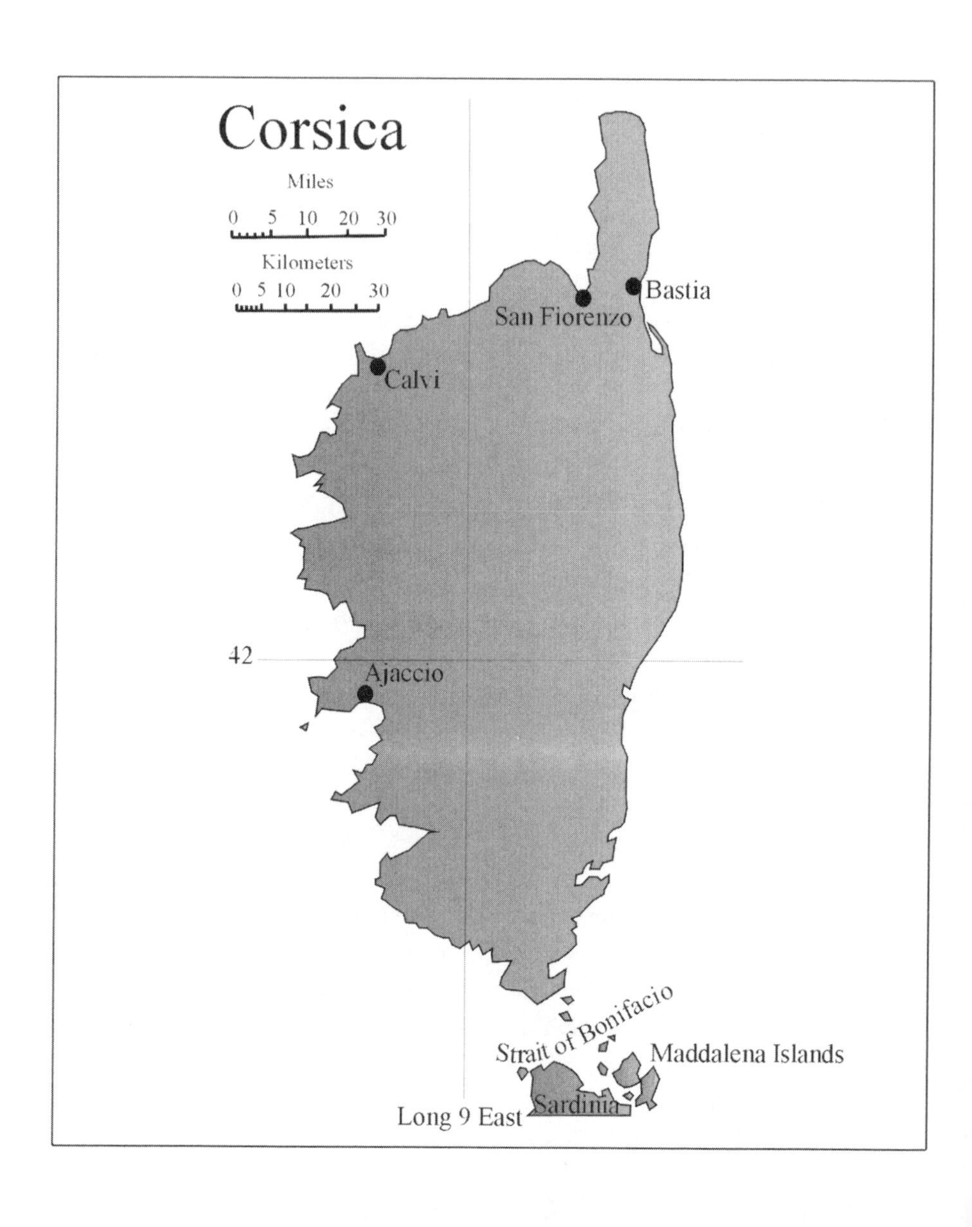
Corsica
Miles
0 5 10 20 30
Kilometers
0 5 10 20 30
Bastia
San Fiorenzo
Calvi
42
Ajaccio
Strait of Bonifacio
Maddalena Islands
Sardinia
Long 9 East

Bastia, however, was now a thorn in Hood's side. Were it to gain reinforcement from France, it might serve as a staging ground for French attempts to take back San Fiorenzo and other parts of Corsica. Hood therefore gave Nelson—who had yet to play a role in the battles for Corsica—a squadron of six frigates and instructions to blockade Bastia closely, ensuring that no food, supplies, or reinforcements got in.

Colonel Moore, meanwhile, moved his troops over the steep mountain pass from San Fiorenzo and on 24 February 1794, began to construct artillery batteries on the heights of Cardo overlooking Bastia. Yet before this could be completed, and an adequate logistics system established to ensure a steady supply of musket balls and cannon shot and shells, the French sallied out of Bastia in strength. Driving irresolute Corsican nationalists from positions supposedly protecting the British batteries, the French took the heights and began to entrench themselves precisely where Moore had hoped to place his guns, prompting General Dundas, an experienced soldier and master artillerist who personally and carefully reconnoitered the front with Moore on three successive days, to order British troops back to safer positions nearer San Fiorenzo. With surprise lost and the French now in control of the heights of Cardo above Bastia, Dundas reasoned that the city could not be attacked from the landward side, at least not before its garrison had been significantly weakened. Any such attempt, he told a grumpy Lord Hood a few days later, would be "most visionary and rash."[31] The naval blockade, Dundas argued, should therefore be intensified with the aim of starving the garrison into submission.[32]

Bad relations between Hood and Dundas dated back to the defense of Toulon, when the admiral had ignored Dundas's advice to prepare for a rapid evacuation.[33] Dundas had been right, and although Hood never admitted it, he paid a heavy price for ignoring the army general. Now, once again, Dundas's reasoning annoyed Hood, even though he had not bothered to reconnoiter Bastia's land or sea approaches himself. Dundas's decision to withdraw from Bastia also angered Hood's devoted subordinate Nelson (in charge of the *Agamemnon* and a frigate squadron), who had studied the sea approaches only.

Nelson's nature had changed little during the years since the siege of Fort San Juan. He still thought virtually anything could be accomplished by courage, audacity, and aggression. After watching Moore's men withdraw on

Dundas's instructions, Nelson insisted in a letter to his wife that he would have taken Bastia by storm, and with as few as five hundred men (a week later he altered his boastful claim to add the *Agamemnon*'s crew and guns).[34] "Army go [*sic*] so slow," he unkindly added, "that seamen think they never mean to get forward; but I dare say they act on a surer principle, although we [navy] seldom fail." Years later in 1800, Nelson would likewise boast to a friend that, were he ever to receive command of an army, he "would only use one word—*advance*—and never say *retreat*."[35]

Although one should not make too much of a single sentence, Nelson's choice of words here is interesting. He clearly did not understand that armies are not always, or even usually, doing something wrong when they withdraw. A withdrawal is merely a transitional phase involving the disengagement of a force from the enemy in accordance with the will of a commander. It usually occurs in order to avoid battle in unfavorable terrain or tactical conditions or to draw the enemy into a battlefield that one has chosen *for* its helpful terrain or tactical conditions. A withdrawal is very different from a retreat, which refers to rearward flight following real or impending defeat at the enemy's hands. Nelson's suggestion that armies were "retreating" if not going forward was either uninformed or most unkind.

No expert soldier, much less an inexpert warrior on land like Nelson, could or would have led five hundred lightly armed men down an open valley (or across an open bay and over a strong seawall, if the attack came from the sea) to take Bastia from a prepared, waiting, strongly armed and fortified opponent. The city's defenders were then believed to include thirteen hundred line troops, the crews of two frigates, and eleven hundred armed Corsicans, all protected behind strong, unbreached fortifications with sixty-two cannons, some mortars, and a good supply of ammunition.[36]

Nelson was still deeply influenced by his experiences of combat at sea. The nature of naval combat resulted partly from the difficulties of reconnaissance, with frigates constantly patrolling in order to find the enemy and then withdrawing to report the enemy's movements to their parent squadron. That squadron would sail as swiftly as it could and strike immediately, before the enemy could disappear in the sea's wide expanses. Sea combat, therefore, usually took the form of what we now term *contact battles*.[37] A contact battle, also known as a *hasty* battle, refers to a clash of forces in which, perhaps because the enemy's whereabouts or movements are not precisely known, one side launches an attack immediately on contact without wasting precious time on preparation. In a contact or hasty attack, preparation time is traded for speed and audacity in order to gain surprise and exploit momentary opportunities.

The alternative to contact battle is now called *deliberate* battle, otherwise known as *methodical* battle.[38] A deliberate battle involves at least one belligerent, and often both, undertaking preplanned offensive or defensive action based on reconnaissance, methodical preparation, adherence to general guiding principles, and the careful deployment of forces in anticipation of the enemy's dispositions. Such battles are more common (and easier to attempt) on land than at sea because land forces generally have better and more frequent intelligence about the enemy's whereabouts, strength, and movements than do sea forces. A deliberate or methodical attack is, and always has been, considered essential against a well-prepared and waiting foe.

Nelson may not ever have learned it, but the offense-defense ratios accepted by most land commanders, already established over centuries of battles and sieges, recommended an attack strength three times greater than the defense's. This was, of course, a rule of thumb that referred to local (not overall) superiority and applied to deliberate battles and sieges only. Many commanders throughout history had executed successful attacks, particularly in fluid and spontaneous contact battles, with much less favorable odds. Few, on the other hand, had done so in deliberate battles, especially when the numerical odds were so heavily stacked *against* them, over such difficult terrain, against a strongly fortified foe, and after the element of surprise had disappeared. (And as it turned out, the first British estimates of Bastia's strength were too low; defenders numbered well over four thousand!)

Even the "young, enterprising" Colonel Moore[39]—who commonly favored the coups de main and actually supported the idea of a land attack on Bastia—agreed with Dundas. He considered his explanations why one should not take place "full of good sense and moderation."[40] In any event, Moore had never planned to storm the city in a spontaneous Nelsonian charge following a quick bombardment. Despite having many more troops available for an attack than Nelson claimed he himself would have attempted it with, Moore had planned a far more careful and methodical attack. His intention was to place batteries of heavy guns on the heights of Cardo. He would then establish a logistics chain that would guarantee him the ability to fire heavy and continuous barrages during successive days and weeks (it would be pointless consuming all munitions in a rapid barrage that could not be immediately exploited).[41] Only after the guns had adequately reduced the fortifications, opened practicable breaches, inflicted bloodshed, and broken morale would he order his troops to advance on Bastia to escalade the walls or enter through the breaches.[42] And none of this would occur unless the navy had successfully blockaded the city.

Admiral Hood and General Dundas argued bitterly over the latter's decision not to launch an immediate attack on Bastia from the land side. The seventy-year-old admiral received strong encouragement from Nelson, who recognized the risks involved but radiated the fullest optimism. Nelson privately confided to the father of one of his protégés that if success at Bastia proved too difficult, Britain would "sooner forgive an officer for attacking an enemy [and failing] than for letting it alone."[43] Yet, he publicly and repeatedly poured scorn upon Bastia's defenses and insisted on the ease of its capture. He also deliberately misrepresented the garrison's strength and morale to Hood by claiming them to be far lower than he knew them to be.[44] Even when he gained unmistakable evidence of Bastia's true garrison strength, Nelson withheld it from the admiral, fearing he might call off the attack. Doing so, he later said privately in weak justification, would have damaged England's, Hood's, and his own honor.[45]

Nelson's duplicity should not be excused (although some hagiographers have done precisely that) no matter how sincere his motive may have been.[46] As a commissioned officer he had a solemn responsibility to give his commander in chief an accurate appraisal, with all facts presented truthfully and none omitted. If asked, Nelson should also have given his honest opinion on how best to proceed. Then, if the admiral chose to adopt a course of action contrary to that advice, which was his prerogative, he alone bore the burden of responsibility. Such is the necessary nature of command. To manipulate the situation appraisal process with misleading information and advice was most improper of Nelson.

Still, Nelson was not Hood's only source of information—Paoli also wanted an attack and deliberately minimized Bastia's resolve—so we must not ascribe the admiral's desire for an immediate attack solely to Nelson's reports. Yet the young firebrand's eagerness for a supposedly quick battle, won at the expense of the army, clearly inflamed Hood's own enthusiasm. Nelson himself thought so, claiming to Fanny: "The expedition is almost a child [of] my own."[47]

Hood would order the attack at his "own risk, with the force at present here," he bluntly told Dundas.[48] As British supreme commander, Hood said, he had the necessary authority. Dundas, who was Hood's equal in terms of rank (if not of age), rejected the admiral's claim of supremacy. He correctly pointed out that they had commanded at Toulon as coequals and that no formal changes to arrangements had since been signaled by the British government. Hood conceded they had been coequals, but maintained that Dundas's formal command authority actually ended with that city's evacuation. He had let Dundas lead troops on Corsica so far and up until this point merely as a courtesy.[49]

General Dundas was naturally disgusted, claiming that if Hood were right, he was dishonest and terribly ungentlemanly in not making this clear much earlier.[50] How could the admiral take away a general's command authority without telling him, while then letting him go on, for three months, thinking he could still command his troops! What courtesy was that? Hood should be ashamed, Dundas stated, requesting the admiral to produce evidence that he had ever received authority from the king to command both the army and the navy in the Mediterranean. (No such proof ever appeared; indeed, it did not exist.)

Hood was out of line. In such situations, the Admiralty insisted that fellow officers in the two services must not attempt ungenerous one-upmanship, but must try to function as equal partners. For instance, in May 1798 the Admiralty provided these emphatic instructions to one captain about to join forces with the army:

> It will also be most essentially necessary to the success of the undertaking that the utmost good temper and spirit of co-operation should subsist between the Army and the Navy. You will therefore use your utmost endeavours to promote such good dispositions, and are to prevent, or remove, as far as can possibly be done, all causes of discontent or dissatisfaction.[51]

What the Admiralty would have made of Hood's behavior, which occurred, of course, well away from the Admiralty's notice, is anyone's guess. One suspects it would have found his behavior's disastrous impact on interservice relations most regrettable.

To Colonel Moore's grave disappointment—he considered Hood's tactical opinions ignorant and his behavior unprofessional—General Dundas announced that he considered the situation so intolerable that he would return to England forthwith on the grounds of ill health.[52] When Hood sensed this coming, he gloated to Nelson: "The General's faculties seem to be palsied [that is, paralyzed]; we must therefore do the best we can. . . . Ever faithfully yours, Hood."[53] Two days before Dundas finally departed on 11 March 1794, Hood unkindly told Nelson that the general had "made up his mind to walk off," thus implying that the general was merely being precious and was, in effect, deserting his post. (No one else of importance saw it thus. Dundas, author of an influential manual on tactics, rose to full general in 1802.)

After Dundas left, Moore and Maj. George Frederick Koehler, an artillery expert (who later became a general), again reconnoitered Bastia and reported to Hood at a war conference attended by senior naval and military officers that the city's defenses were indeed too strong to be attacked.[54] Moore recalled that a lot of "foolish conversation" occurred at this 20 March meeting. All army officers argued that no attempt should be made, except for a young sub-lieutenant who was out of his depth and apparently found Nelson's optimism irresistible.[55] All sea officers, few of whom had walked the ground (as Nelson had, although as a sea officer he had never been trained to assess enemy fortifications and strengths), insisted that it should. Admiral Hood overruled the army, to Moore's distress, and ordered a land attack from the north. He bullied Dundas's immediate subordinate, Brig. Gen. Abraham D'Aubant, who was (according to one observer) "totally unequal to the position," into giving him troops, guns, and stores.[56]

Nelson continued to ridicule the army, accusing it privately of excessive timidity. He also asked Hood for command of the attack force, or at least its naval elements. Delighted by Nelson's enthusiasm and aggression, Hood gladly acceded to his request. On 4 April 1794, Captain Nelson and Lt. Col. William Villettes, an army officer chosen by Hood as nominal tactical co-commander, strode ashore three miles north of Bastia at the head of a force comprising over twelve hundred soldiers and seamen.

Siege Warfare: Bastia

While Villettes's soldiers marched south toward Bastia, Nelson and his seamen struggled with the guns. There were at least sixteen: one little 4-pounder, three 12-pounders, and five 24-pounders from the *Agamemnon,* as well as a howitzer, four mortars, and some carronades.[57] These were all awkward to land and, in the total absence of pack and draft animals, backbreaking to pull into position close to the city's defenses. Teams of seamen had to clear ground upon which the hastily made sledges carrying the guns would travel. (Unlike army field pieces, with large-spoked wheels, naval guns had special truck carriages with small wheels for recoil on ships' smooth decks. These carriages made the guns remarkably hard to move over almost all terrain.) They then had to pull the guns up steep hills and ridges and across ravines by means of complicated tackles using 2.5-inch and 3.5-inch diameter ropes and heavy pulleys, all ordered from Commodore Linzee in anticipation of the exhausting work.[58] Aside from the gun barrels, of course, the seamen also had to haul the planks, balks (heavy timber beams), gabions (baskets of stones used in building fortifications), sandbags, and artificers' tools. They then had to construct the batteries from these materials, as well as dig trenches and raise breastworks. Nelson was full of admiration, writing home: "It is a very hard service for my poor seamen, dragging guns up such heights as are scarcely credible."[59] Nelson and Villettes cooperated well, given the bitter enmity now existing between the navy and the army and the scorn Nelson privately felt for Villettes's superiors.

Hood had predicted that the batteries would be complete within two days and bring victory within ten more. This assumption was wildly optimistic. A full week passed before the batteries were able to open fire on Bastia on 11 April. And the results, although seemingly impressive as the valleys echoed with great booms, did not include the predicted weakening of the enemy's fortifications or morale. After almost two weeks of shelling, which consumed over five hundred barrels of powder and six thousand roundshot and three thousand shells,[60] all rolled or carried forward by exhausted seamen, Colonel Moore jotted in his diary: "The cannonade and bombardment of Bastia still continues. We have not gained an inch."[61] Hood's boasts that the garrison would give up as soon as they suffered gunfire had clearly proven illusory, Moore scornfully added.

D'Aubant had come to share Moore's sentiments regarding Hood's bullying and the attack's obvious uselessness. So D'Aubant refused to provide further troops until the issue could be decided by a new general reportedly coming out from England. Interservice relations became so bad that on one occasion after Hood unsuccessfully requested two 10-inch howitzers, he went ahead and took them anyway.[62] D'Aubant watched in disbelief but could do nothing. Nelson, incapable of gaining ground, and now aware that Bastia was a far tougher nut to crack than he had earlier boasted, took Hood's side in the squabble. Nelson unfairly blamed D'Aubant for refusing to reinforce the attack that was so clearly petering out.[63] If the siege continued for much longer, he wrote to Fanny, D'Aubant and his staff "deserve to be shot for allowing a handful of brave men to be on service unsupported."[64]

As it happened, D'Aubant eventually caved in to pressure and authorized reinforcements to join the attack. Nelson, fed up with sleeping in a tent next to his batteries and not being able to see any tangible progress, but still longing to lead an assault, looked forward to a real clash of arms occurring.

Before D'Aubant's troops could join the attack, however, the Bastia garrison and citizen militia suddenly surrendered on 19 May, to the great surprise of almost all onlookers. The defending forces, and the city's ordinary citizens, were simply too hungry to continue resisting. Hood, Nelson, and their naval colleagues claimed victory; and in truth the victory was theirs, but *not* because of their land attack. That had degenerated into a very costly and ineffective siege. The French garrison, whose fortifications were largely undamaged and who still had good stocks of ammunition, surrendered because the naval blockade starved the city into submission. Moore scornfully noted in his diary after news of the surrender first reached him: "Had the fleet block[ad]ed the port without landing a man or firing a shot the place would at this instant have been equally near a surrender."[65] The British Army's semi-official account is just as critical of the navy, but no less accurate:

> The fact remains . . . that Bastia was reduced by starvation only, and that Hood's operations ashore, [encouraged and co-led by Nelson and] undertaken directly against the advice of Moore and Koehler, were futile and absurd, wasting much valuable ammunition, which was none too plentiful, and not hastening the fall of the place by one day.[66]

Nonetheless, Hood remained convinced that the victory was his, and privately praised Nelson for his contribution to Bastia's surrender. Nelson's seamen had demonstrated "indefatigable zeal and exertion" as they carried out very difficult and exhausting tasks.[67] Hood would remember their efforts to the end of his life. Nelson received this praise with undisguised satisfaction. Apparently forgetting his earlier boastful predictions of a short, sharp, and relatively easy battle, and the subsequent reality of a hard, lengthy, and ineffective siege, he told his wife: "Lord Hood has gained the greatest credit for his perseverance and I dare say that he will not forget that it was due to myself in a great measure this glorious expedition was undertaken. . . . When I reflect on what we have achieved I am all astonishment."[68]

Siege Warfare: Calvi

No sooner had Bastia surrendered than Hood ordered D'Aubant to provide him with two regiments. With these Hood ("playing soldier" again, to quote the army's semi-official chronicle) planned to take Calvi, another French-occupied city.[69] Before further interservice squabbles could occur, however, a new army commander arrived in-theater on 25 May with a commission as the lieutenant general of Corsica. This gifted officer, the Honorable Charles Stuart, was thus empowered to command his troops without interference from Hood. Stuart was sensitive and wise, though, and realized that his first task should be repairing the damage done to interservice relations. He went to great lengths, by means of calm mediation and words of encouragement to all involved persons, to smooth ruffled feathers. It worked, and with D'Aubant departing for Leghorn, planning for the capture of Calvi began in a spirit of interservice cooperation.

Calvi would be harder to take than Bastia, as Stuart realized after his first reconnaissance missions. It sat on a rocky headland, with the walls of its citadel and the cliff beneath rising more than two hundred feet sheer from the sea.[70] Even the parts of the city lower down around the harbor were heavily fortified and covered with outer defenses. Unfortunately for any attackers, the only possible site from which to launch a direct amphibious assault was the bay immediately beneath the citadel under the sweep of its powerful guns. Stuart therefore planned an overland attack to be made by a force landing on the island's west coast and advancing on foot over steep hills. French outer forts would be besieged and taken, before the city and its citadel would likewise be attacked and then, once practicable breaches had been opened, assaulted by the army.

On 17 June 1794, the joint army-navy force of sixteen transports, victualers, and storeships, protected by Nelson's *Agamemnon* and two smaller ships, anchored off Corsica's west coast. On 18 June Nelson and Stuart conducted a predawn reconnaissance of that rocky coastline in search of suitable landing sites within reasonable distance of Calvi.[71] They found only a tiny inlet known as Porto Agro, three and a half miles southwest of Calvi. It was hardly ideal, with jagged rocks beneath the surface and an unhelpful sea breeze. Yet it would have to do. There was nowhere else. So landing began at once, with the first of 250 seamen going ashore with Nelson to manage the unloading of guns—three 18-pounders, then four 26-pounders (the first of twenty), and six field guns[72]—followed by the 2,300 troops themselves.[73]

Heavy swells and poor weather made the landing difficult. Then Nelson's seamen proved inadequate for the difficult task of hauling the heavy guns up the rock faces and steep hillsides toward the heights overlooking Calvi. Three hundred soldiers had to assist, much to Moore's chagrin and Nelson's disappointment.[74] It could not be helped; it took two hundred or so men to haul *each* of the 26-pounders (the barrel of which weighed over two tons) up to the heights, and that did not include the powder, shot, and shells.[75] Still, the two services generally cooperated harmoniously at this tactical level, with Nelson and his army counterparts struggling with their task as a team of equals. Nelson even liked General Stuart, at least initially, and worked hard to meet his expectations. He was delighted when Stuart consulted him on gunnery matters,[76] and closed one letter to the general with this sincere sentence: "I can only assure you that every exertion of mine shall be made to comply with your wishes on every occasion, for I am, with the highest esteem, your most faithful servant, Horatio Nelson."[77] This high esteem would not last long.

Nelson accepted that Calvi would have to be taken by siege, not by storm, and this time made no appeals to lead an assault. Perhaps he had learned from his experiences at Bastia after all. He nonetheless believed since British forces were constructing their batteries out of French view, they might be able to exploit surprise, raining down such an unexpectedly heavy bombardment that the French would soon sue for peace.[78] Even after French forces sallied out of Calvi on 27 June, killing two dozen Corsicans before being driven back, he remained optimistic of a short siege.

However, sieges very rarely involve surprise. It is remarkably difficult to wage this type of artillery battle without undertaking exhausting and careful preparatory efforts, most carried out at night but all within the enemy's hearing distance, if not view. Nelson's optimism began to fade by early July when he realized the army's attack would proceed, even more than it had at Bastia, according to the established principles of siege craft.

These principles ensured that the following conditions would be met: batteries would be positioned scientifically at the exact distance from enemy batteries or the sections of fortification to be attacked so as to maximize the strength of the different caliber guns; these batteries would be dug into the ground and supported by sufficiently strong and carefully constructed platforms; breastworks would be raised and embrasures dug in order to protect the artillerists and the guns from counter-battery fire; camps would be created to ensure the shelter, welfare, and effective work shift rotation of artillerists and soldiers; a logistics system would be established to keep artillerists fed and healthy as well as continuously stocked with essential supplies. These supplies included the 1,090 barrels of gunpowder and the 11,275 shot and 2,751 shells the artillery would eventually consume.[79] All of the ordnance had to be rolled, pulled, and carried over miles of rough ground—and entirely without the benefit of the oxen and other draft animals the army would ordinarily employ.

Impatient for victory, Nelson found this too time-consuming and, although he never let his efforts flag, even a bit boring. As a warrior who thrived on the verve and pathos of fluid contact battles, he also found this sort of preparation painfully methodical. As he complained to Lord Hood on 5 July: "A happy degree of irregularity I can't help thinking is sometimes better than all this *regularity*."[80] Indeed, the methodical nature of siege craft had little in common with an infantry or combined-arms battle. And it bore absolutely no resemblance to a sea battle, where ample stocks of powder and cannonballs were kept (as were the crews' food and drink) in close proximity to the guns themselves, and where spontaneity and passion dominated. This was the world that Nelson sorely missed and could not wait to return to.

The contest for Calvi was not one-sided, of course. The French outer forts exchanged fire with the British batteries, preventing the advance and slowing down their rate of fire. As Nelson jotted in his journal on 8 July 1794:

> Throughout the whole of the 8th, both sides had kept up a constant and heavy fire. They [the French] totally destroyed two of our twenty-four-pounders, greatly damaged a twenty-six-pounder, and shook our works

> very much. One of their shells burst in the centre of our battery, amongst the General, myself, and at least one hundred persons, and blew up our battery magazine, but, wonderful to say, not a man was much hurt.[81]

This entry reveals the scale of the siege. The operation of this battery alone required a hundred artillerists and laborers, excluding those involved in bringing munitions forward. And the British had four batteries of this size.[82] This journal entry also reveals the grave perils of artillery duels. Stuart, Nelson, and their gunners were extremely lucky to survive this one French shelling. Nelson proved equally lucky four days later, on 12 July 1794, when a French ball crashed down in front of him, spraying gravel and dirt like deadly shrapnel. Smashed to the ground, he arose to find himself covered in blood and unable to see much with his right eye.[83] The sight dimmed further and never returned, leaving him virtually blind in that eye.

On 18 July, General Stuart finally felt that the breach in the mighty Fort Mozzello was practicable and ordered an infantry assault to begin one hour before the following dawn. Colonel Moore and Col. James Wemyss led the attack, which managed, despite deadly musket and cannon counterfire, to take the fort and the neighboring Fountain Battery. Nelson had no role in this assault, much to his disappointment. Instead, he and his seamen continued pulling guns forward and constructing new batteries, which could now pound the citadel itself. They opened fire on 31 July, after the French garrison commander had refused—albeit with great courtesy and obvious concern for the welfare of his men—Stuart's summons to surrender. He would fight on, the Frenchman said, as long as there was any chance of getting reinforcements and supplies. Yet if none arrived by 10 August he would surrender.[84]

In the meantime, on 20 July Stuart asked Nelson to take his ships right up to Calvi, so they could bombard its defenses with their guns. Nelson refused to lay his "wood before walls," as this type of direct shore bombardment was casually called. He explained to the general the grave probability of being attacked with red-hot shot (cannonballs heated intensely in furnaces before being fired). These were naturally extremely dangerous to ships made of wood and carrying tons of gunpowder. In any event, attacking forts from the sea was not generally an effective use of ships' guns.[85] Seaborne guns were always less accurate than those fired from land batteries. Stuart accepted this advice without protest.

Typical of sieges throughout history, the greatest threat to the besiegers' lives was probably not counterfire, but the effects of physical exhaustion and adverse weather. Unable to gain shelter from the intense July heat (even Moore, Stuart, and Nelson slept at their batteries), with water supplies always low, and with physical exertions constant, the besiegers began falling sick at an alarming rate. On 30 July Nelson lamented to Sir Gilbert Elliot (the first earl of Minto and Hugh Elliot's older brother): "We have much more to dread from the climate than from the fire of the enemy. . . . They know their climate, that it is an enemy we can never conquer; for if the siege is prolonged one more week, half this army will be sick."[86] That same day Moore noted in his diary: "The men and officers fall ill daily; considerably more than a third of our force are in the sick report."[87] On 4 August Moore was forced to write that over *two*-thirds (approximately fifteen hundred from the original twenty-three hundred[88]) were now in the field hospitals and hospital ship, and that another week would finish off the siege.

Thankfully for the British, who were indeed about to lift their siege, the French garrison kept its word and surrendered on 10 August 1794. Having received no help from France, it had run out of ammunition and could no longer fight on. Six hundred French soldiers marched out of the citadel in orderly fashion past three hundred enfeebled British troops. Some of Moore's regiments had only twenty men fit for duty.[89] This outcome was not the most glorious close to the Corsican campaign, but at least it was over and in Britain's favor. Nelson could return to sea.

Even before Calvi surrendered, Hood and Nelson were once again feeding each other's dislike of the army and its methods. Both felt slighted that Hood had not been permitted at the end of June to issue a surrender summons to the Calvi garrison. Stuart had told Hood that it was Stuart's siege, not Hood's, and that he should mind his business.[90] Then, as the siege progressed through July, Hood and Nelson had sent each other letters cryptically denigrating Moore and even Stuart (a basically good but weak and deceived man, they thought).[91] They became secretive in what they said and did because the army purportedly still possessed the same antinaval spirit (or the "Saint Fiorenzo leaven," to quote Hood) as it had under Dundas.[92] Now, with Calvi in British hands, the admiral and the captain claimed that Stuart had been too lenient on the French after the capture of the outer forts on 19 July and should not have waited, with only sporadic bombardment continuing, for their surrender on 10 August.[93] Stuart should have threatened the city's immediate destruction unless the garrison surrendered forthwith.

Angered that he was not mentioned in the deed of surrender, Hood wrote home in what one scholar has recently called "a mean spirit, blaming Stuart for prolonging the siege."[94] Hood then requested from Stuart an unrealistically high number of soldiers to serve on his ships as seamen and marines.[95] The general provided what he could, even though the admiral suspected, and later complained to the Admiralty, that he was selfishly holding back many. Hood then promptly removed his ships from Calvi, leaving the disbelieving army incapable of dismantling its batteries and removing its stores.[96]

Nelson's dislike of the army reached new heights in the weeks and months following Calvi's surrender. He realized to his dismay that he would receive little public recognition for his role in the battles for Corsica. He had been hopeful of favorable mentions in Hood's and Stuart's dispatches, especially after Hood had failed to give Nelson adequate public praise in Hood's earlier dispatch on Bastia's fall. Surely this unintended snub would not be repeated, he reasoned. He was devastated, therefore, eventually to learn that, while Hood had given him faint praise, Stuart barely mentioned him at all in his dispatch published in the *London Gazette*. Nelson fumed to Fanny that Stuart's hatred of Hood had doubtless spilled over to include him.[97] This hatred was, he insisted, "founded on the most diabolical principles," because the army was jealous of the navy's seizure of Bastia and could not fault Hood's cooperation at Calvi. (As noted, it could and did!) In truth, Nelson claimed, all the heavy work and the artillery firing was done by seamen under the command of sea officers. "This is my crime," he added, "and I suppose the Scotchman [Stuart] will do all he can to injure me, so far as not saying anything for me, or any other way I defy his malice." He could forgive but not forget those who had ignored his invaluable contribution, he told William Suckling. Were it not for his overwhelming sense of duty, he might well retire from naval service in disgust at his inability to get fair recompense for his efforts.[98]

Nelson was being overly sensitive and dramatic. He and his seamen had certainly toiled exhaustingly during the battles for Corsica. Even the army's semiofficial history records that, particularly at Calvi, "Nelson's zeal and industry were indefatigable."[99] Yet, while zeal and industry are marvelous traits, they seldom earn significant mentions in dispatches unless accompanied by acts of heroism, invaluable service of some kind, or a key role in successful tactical planning. Nelson and his crews suffered fatigue, sickness, and combat casualties but carried out no heroic actions. They landed the army successfully and then provided much of the back-breaking labor needed to get guns in place and build and man the batteries, but they did this in concert with army laborers and artillerists. With the exception of Bastia, they had little input into the planning of the battles. They acted under army instructions with no opportunity for tactical moves of their own. And the battle for Bastia, which Hood and Nelson did plan together, against the advice of the army, proved costly and ineffective. The naval blockade deserves almost sole credit for that city's surrender.

It seems clear, then, that Nelson's experiences on Corsica during 1794 did little to enhance either his understanding of the principles of land warfare, which remained poor, or his respect for the British army, which remained low. His constant advocacy of hasty attacks, in which audacity and courage would supposedly overcome even great numerical and material inferiority, was counterproductive and, in the case of Bastia, misplaced and costly.

To be fair, however, he worked under the command of a very strong-minded admiral, whom he adored and worked zealously to please, and who had an equally poor understanding of the limitations, capabilities, principles, and tactics of armies. Inspired by Hood's intensity and larger-than-life personality, Nelson soaked up the aging admiral's preconceptions and biases, which generally matched Nelson's own, and even sometimes copied his jealous and petty behavior.

Moreover, in Corsica Nelson had no opportunities to see the army undertake any fluid battles of maneuver. None occurred. In such battles the combat traits he admired—courage, audacity, and aggression—tended to be far more visible and influential than they were in the siege craft he found himself undertaking on Corsica. Had he been present at any of Stuart's or Moore's other battles against the French or Spanish, he may well have formed a different opinion of the army's and its commanders' abilities. But helping Moore and his colleagues to pound fortresses and field fortifications in methodical attacks was hardly likely to inspire a firebrand like Nelson, whose experiences, training, and temperament left him unsuited to the intensive labor, careful method, and slow pace of army siege craft. One cannot, therefore, blame Nelson for not being more positive about the army's activities. He was a product of his naval background and a prisoner of his excitable, passionate personality, which combined to make him an unhappy, uninformed, and impatient soldier, as well as an energetic, expert, and stunningly successful seaman. Thankfully, he returned to sea later in 1794.

Continuing Misconceptions

For the next two and a half years, Nelson had less to do with British or allied armies and their operations. He watched with frustration as an Austrian army led by Gen. Bernard de Vins accomplished little in the Italian Riviera. He privately told friends and family that the Austrians lacked aggression and constantly made excuses not to attack.[100] For example, on 5 September 1795 he wrote to his wife: "I am not quite so well pleased as I expected with this army, who are slow beyond all description . . . In short, I can hardly believe he [de Vins] means to go any further this winter."[101] Nelson was equally unimpressed by Gen. Johann Peter Beaulieu, who later replaced de Vins. Nelson complained that Beaulieu was also passive and unimaginative, and that his failure to conform to the planned timing of an attack robbed him (Nelson) of the opportunity to bombard the French troops as they withdrew along the Riviera coast.[102]

In July and September 1796, Nelson cooperated well with the army in the taking of Elba and Capraia, both of which capitulated to British demands without battle. Although the art of war is undoubtedly to win without having to fight, the absence of battle on those occasions robbed historians of further examples of Nelson fighting ashore. Nothing much can be said of those operations, except that they were very small in scale, there were no competing egos, and, perhaps consequently, Nelson interacted well and happily with the army. Certainly the operations provide no evidence of what one leading Nelson historian mistakenly called "Nelson's mastery of the art of joint operations."[103]

Then, later in 1796 Sir John Jervis ordered him to evacuate all British troops on Corsica. Nelson, who had spent most of 1794 struggling to take the island from the French, naturally found the mission bitterly distasteful. If only allied armies had beaten the French somewhere on land, he thought, the Spanish might not have changed sides and the Mediterranean would still be safe for British warships.

Aside from these events, however, he spent two years happily at sea, which proved a productive period. He first earned honor for his role in naval engagements in March and July 1795 and then fame and acclamation for his courageous, masterful, and decisive action during the Battle of Cape St. Vincent in February 1797. No longer Captain Nelson, a respected but little known officer, he became Rear Adm. Sir Horatio Nelson, knight of the Order of the Bath and a national hero.

His next involvement with amphibious or land operations came in the form of his ill-fated assault in July 1797 on the town of Santa Cruz de Tenerife in the Canary Islands. The key elements of the battle have been recounted already in this book and need no repetition. Yet there are aspects that deserve analysis pertaining to Nelson's relationship with the army, as well as to his attempt to plan and direct operations ashore.

Both the earl of St. Vincent (as Sir John Jervis soon officially became) and Nelson wanted the army's participation in what they originally intended as a large-scale amphibious assault made *by soldiers* assisted by naval marines. It would be a relatively easy operation, guaranteed not to fail. This, at least, is what Nelson told St. Vincent in a letter very reminiscent of those he had sent to Hood before the siege of Bastia.[104] The soldiers and their cannons, mortars, and provisions could, Nelson wrote, be landed from troop transport vessels near Santa Cruz in a single day. The combined army-marine force could then win the battle in three days, "probably much less."[105]

Nelson certainly did not intend a prolonged siege like those of Bastia and Calvi. Instead he planned a night landing of all forces followed by a fast and powerful surprise attack. The army would cut off Santa Cruz's water supply, then pound it with cannon fire from the heights of Jurada behind the town before storming it with infantry and marines. Surprise and speed would play key roles, as would confidence and coercion. Should Santa Cruz refuse an appeal to surrender and instead resist militarily, its "utter destruction" by artillery fire from the heights would be carried out.

Nelson encouraged St. Vincent to seek troops from Lt. Gen. John de Burgh, whom Nelson knew and respected (and from whom he had borrowed three hundred troops for his seizure of Capraia). This time de Burgh might well contribute his thirty-seven hundred troops on Elba to the planned attack on Santa Cruz.[106] By a happy coincidence, their cannons, mortars, and equipment were already embarked on naval transports at Porto Ferraio (in anticipation of their evacuation to a safer location). If de Burgh declined, Gen.

Charles O'Hara, governor of Gibraltar, should be asked for some of his troops. They would only be needed for a few weeks.

Gaining these troops might take a lot of persuasive effort, Nelson advised St. Vincent. Nelson had learned "from experience" (presumably on Corsica) that soldiers "have not the same boldness in undertaking a political measure that we have; we look to the benefit of our Country, and risk our fame every day to serve her; a Soldier obeys his orders and no more." Clearly Nelson's dislike of the army's way of war had not decreased. Nor had his optimism. His plan could not miscarry, he boasted to St. Vincent. It would be so easy that (if it came to it) even a thousand troops and some marines could carry it out.

The plan—at least in this form, before its ruination by Nelson himself—had much merit. It involved an indirect approach, with the attack force bypassing the strength of Santa Cruz's strong frontal fortifications to occupy the heights of Jurada thought to be immediately behind the town, upon which light gun batteries would be quickly assembled. With positional advantage gained, the attackers would launch an aggressive artillery barrage to break the morale and cohesion of the town's surprised garrison and citizens. Finding the British behind them and well situated and armed, the besieged Spanish could not easily mount a defense, much less an offense, against a foe thus situated. With resistance seemingly futile and with extra coercive weight coming from the sight of Nelson's squadron of warships, visible on the horizon, the enemy might indeed accept his demand for their surrender.

Successful execution of the plan rested on four critical elements: (a) maneuvering quickly and effectively to gain positional advantage over the enemy; (b) doing so without the enemy's knowledge or ability to interfere; (c) having the hundreds of men needed to pull the 18-pounders and their platforms and ammunition into position; and (d) having a sufficient show of manpower and firepower to intimidate the Spanish into submission. Unfortunately, when Nelson eventually undertook the attack in July 1797, none of these critical elements could be obtained. How could they? They relied on the army's contribution of troops and firepower, and both de Burgh and O'Hara had declined to contribute anything to what they obviously considered an excessively risky venture.

Nelson should have called off the attack when it became clear that the army would not contribute de Burgh's thirty-seven hundred soldiers, or even the thousand that would enable Nelson and his marines and seamen to take Santa Cruz, as he had boasted earlier. Instead, he decided to conduct the attack without *any* soldiers, using only his available marines (around 250) and his unsuited seamen (750).[107] And instead of the force being commanded ashore by an army officer, Nelson decided that Captain Troubridge, whom he flippantly called "General Troubridge" when describing the plan to St. Vincent, would lead the attack.[108] Troubridge would gain the assistance of Capt. Thomas Oldfield, an experienced marine officer. (Oldfield, incidentally, died tragically but heroically during Napoleon's siege of Acre in 1799.)

In this era, marines were trained in musketry, especially *parapet fire* in line along the side of a ship. They also received rudimentary close order drills for dealing with tactical firefights ashore.[109] This training made them almost as useful ashore as regular marching foot. Ordinary seamen were a very different story. Many—not necessarily most—received basic training in how to clean, load, aim, and fire muskets; but very few ever received training (let alone practice) in volley fire. And while soldiers and many marines knew, after extensive periods of drill, how to perform close order maneuvers in unison—quickly and in response to crisp spoken commands or orders relayed by drums—almost no seamen had these essential combat skills.[110]

Nelson did attempt to improve his seamen's abilities by instituting a week or so of twice-daily training in musketry.[111] This short period of training was better than nothing, but not by much. And many seamen were to go ashore with the weapons they used during sea boardings: cutlasses, pikes, and pole-axes, which were of limited use in a firefight. Moreover, although marines in that period could fire ships' guns at close-range targets, the sea soldiers were not trained in artillery.[112]

Without a large body of regular army troops and their firepower and artillery expertise, as well as the reliable on-the-spot tactical advice of experienced army officers, the attack had little chance of success. Yet, once again, Nelson's belief in Providence and his own combat superiority, combined with his insistence that courage, audacity, and aggression were a battle-winning formula, interfered with his ability to read omens. He certainly did not recognize the looming specter of defeat.

He did devote considerable attention to the *techniques* his men should employ. He ordered the construction of special scaling ladders (which show only that he planned an escalade's inclusion—a remarkably risky venture without massive close fire support), as well as the construction of siege platforms and gun sleds (although he had too few men to haul the platforms or the guns). He directed that boats be kept together by towing one another to ensure that they arrived on shore at the same time. He ordered oars to be muffled to enhance surprise. And he had as many seamen as possible dress in spare marine uniforms so as to create an illusion that his attack force was made up of regular soldiers.

Yet even the best battle craft—the techniques and procedures employed by combatants—is wasted if not matched by equally good *tactics.* These small, seemingly important techniques contribute far less to a battle's outcome than many people think. The outcome primarily rests upon the soundness of the overall plan and the tactics, the presence of their supporting components, the quality of the battle drill, and the experience of the fighters. These were all lacking at Santa Cruz.

Earl St. Vincent deserves some responsibility for the fiasco that followed Nelson's decision to attempt the attack with a total assault strength of only a thousand seamen and marines and a few small field guns (3- and 4-pounders, and one or two 18-pounders).[113] St. Vincent surely placed little importance on the prospect of coastal bombardment by Nelson's ships. To bring them within range of Santa Cruz's fortresses, Nelson would have had to sail and probably anchor them under the sweep of Spanish shore guns, which would create a dreadful risk of destruction by red-hot shot, especially to the larger ships. (Nelson must also have realized this; he had himself refused to sail his ships within range of Calvi's forts three years earlier.) In any event, the British ships could not have coordinated their gunfire easily, if at all, with the activities of those fighting ashore. Nelson's sole bomb vessel and smaller frigates might be of some use, but, again, not that much.

No, St. Vincent should have called off the attack, realizing that the assembled force was woefully insufficient to obtain the objectives. Instead, on 14 July 1897, he authorized Nelson to proceed to Santa Cruz with the utmost speed and to take the town by a "sudden and vigorous assault."[114] These words reveal that St. Vincent also considered that the keys to the attack's success were Nelson's courage and the plan's audacity and aggression. The words are also surprising. As a vice admiral in 1794, Jervis (as he was before the Battle of Cape St. Vincent) had led the large expedition to the West Indies that captured Guadeloupe, Martinique, and other islands. These feats were not accomplished by seamen, but by strong contingents of regular army troops (grenadiers and light infantry) under Lt. Gen. Sir Charles Grey.[115] Jervis landed six thousand troops on the weakly garrisoned Martinique alone, which nonetheless took over six weeks to secure.[116]

Indeed, the July 1797 attack on Santa Cruz proved a costly failure. To compensate for having a *far* weaker attack force than originally intended, Nelson and his captains modified their plan, deciding against an attempt to place guns on the heights behind Santa Cruz. They would instead land on a weakly held beach and then storm Santa Cruz's Paso Alto defenses directly in an infantry-style attack with limited fire support.[117] This change was not a wise departure from the original plan, sacrificing as it did any chance of gaining positional advantage over the defenders or employing heavy firepower ashore.

As it happened, high winds and adverse currents prevented the assailing force under "General" Troubridge from reaching their landing points under the cover of darkness. The men were quickly spotted, thus denying them the surprise upon which the entire plan so critically depended.[118] The Spanish then defended their harbor so fiercely that the landing parties had to withdraw to their ships.

After conferring with Nelson, Troubridge agreed that the heights of Jurada behind Santa Cruz should now be taken, as per the original plan. (This, of course, was supposed to involve numerous artillery pieces bombarding or threatening the town, an impossibility after the army chose not to provide its troops and field guns.) Troubridge duly led his force ashore again, this time with only one little 3-pounder "grasshopper" field gun and a few artillerists to aim and fire it. But a new setback emerged; the heights were not directly behind Santa Cruz at all, as Troubridge and his exhausted and sun-scorched party soon learned when they reached the ridgeline. The heights were separated from Santa Cruz by a deep valley that could not be seen from ships outside the harbor (a typical intelligence failure of naval planners, few of whom mastered the complexities of terrain). Taking Santa Cruz from that direction was out of the question, and the one little grasshopper gun would be worthless.

Troubridge withdrew his force again. On 23 July, Nelson held a council of war to consider the options.[119] Given that all surprise had been lost, that positional advantage over the enemy could not be attained, and that their force had proven so weak in men and matériel that it could not even gain local superiority, they should certainly have called off the attack. This is a logical deduction, but then logic seldom prevails over the passions in combat situations. The failure of two assaults under his general command was such a wound to his and his captains' sense of honor that—ignoring his own best instincts and even common sense—Nelson agreed one further attack should be attempted.

This he did in grand style, making a brave but ill-considered departure from his usual maneuverist instincts by launching a frontal attack on Santa Cruz's *strongest* point: the mole head and the central castle of San Cristobal.[120] He led the attack himself, trusting in Providence and the moral effects of audacity and aggression. This assault should never have taken place. Naive efforts to obtain surprise failed, and the attack collapsed almost immediately under the interlocking killing fields of Spanish guns, which cut down many British seamen in merciless sprays of iron. One party that did make it ashore found itself cut off, outnumbered, and helpless. Beaten and bloodied, Nelson called off the attack. He had lost the Battle of Santa Cruz, many men, and his right arm.

Explaining the disastrous attack is not difficult. Nelson had come to consider himself as good at battle on land at he was at sea. Shortly before Calvi's surrender in August 1794, he had written to Fanny: "This day I have been four months landed . . . and feel almost qualified to pass my examination as a besieging general."[121] In April 1796 he had written to John Trevor, British minister at Turin: "I am vain enough to think I could command on shore as well as some of the Generals I have heard of."[122] Nelson's mistaken self-confidence (in this sphere, anyway) cost him and his squadron dearly.

To his credit, Nelson accepted responsibility for the defeat and blamed no one else, not even the army commanders who had refused to contribute troops. That is not to say that Nelson's respect for the army increased. It did not. He remained ever after convinced of the navy's superiority over the army in terms of its strategic importance, professionalism, fighting spirit, and effectiveness. Thankfully for all concerned, he managed to keep this view discreetly limited to comments peppered throughout his private correspondence. It caused no public difficulties. Indeed, Nelson outwardly maintained harmonious relations with army officers stationed in his successive combat theaters. He even initially got on well with Gen. Karl Mack von Leiberich after the

Austrian arrived in Naples in October 1798 to take command of the Neapolitan army. Mack, who proved hopeless and caused a disaster in Naples, eventually disappointed Nelson in the same way as had every other army leader. Unable to understand army tactics and operational art and frustrated by the methodical nature of the land battles he witnessed, Nelson came to see all senior army officers as excessively cautious, timid, and slavishly bound to conservative tactics.

His own disaster at Santa Cruz appears to have taught him two things, however. First, marines and seamen must be better trained to function ashore as soldiers. The former group should, for example, have their own artillery. Trying to get army artillerists to cooperate with the navy and accept orders from sea officers was simply not working. Marine artillery should therefore be raised and kept under the Admiralty's control, be readily available for amphibious missions, and serve under the direction of fleet or squadron commanders. As it happened, Nelson's recommendations to the Admiralty received a very positive reception. In 1804 it raised its own Royal Marine Artillery companies.[123]

Second, and most important, Nelson finally learned that he may have possessed the right combat traits, but his training, experience, and career progression had robbed him of the opportunities to develop a mastery of land warfare. Whereas in 1796 he felt able to boast that he could command as well ashore as he could at sea, after Santa Cruz he became less convinced. In December 1799 he explained to William Wyndham, British minister at Florence, why he [Nelson] could not give Austrian general Peter Palffy the particular assistance he had asked for. As much as he appreciated Palffy's suggestions and flattery, Nelson told Wyndham that "as I cannot be a judge of the movement of an Army, so he cannot of that of a Fleet."[124] For Nelson, this was a significant concession.

Conclusions

As a soldier Nelson was out of his natural environment. He never fully appreciated that battles on land, even small ones, are fought according to different principles and tactics to those fought at sea. His belief that courage, audacity, and aggression are unparalleled force enhancers is certainly true, as is the fact that he possessed those traits in abundance and could inspire them in others. Yet these alone could not compensate for his unfamiliarity with the principles of land warfare and the tactics and techniques pertaining to battle ashore. His own ideas on the way battles should unfold were commonly inadequate and potentially dangerous. We must be grateful that army officers overruled or ignored him in Nicaragua and on Corsica, and that his losses at Santa Cruz, his first attempt at planning and directing a joint operation by himself, were not worse than they were.

Effective and harmonious jointness is difficult to achieve even today, when joint training provides necessary practice and experience, and joint doctrine publications by the dozen spell out the roles and responsibilities of joint staff officers and warriors. So we must not expect too much of a fighting seaman from the eighteenth century when no training or doctrine specifically for amphibious operations existed, and when his own force, the Royal Navy, was widely held to be the "senior service." Yet Nelson's personality worked against his performing better on land than he did. His relations with army officers were seldom collegial, often tense, and sometimes bad and counterproductive. Although he did better on some occasions than others, in general he was not a good joint warrior.

In Nelson's defense it must be noted that he had been trained since the age of twelve in naval tactics and methods, but not in the business of soldiering. He read prodigiously, as did Napoleon, but naturally limited his study of warfare, again like his counterpart, to the art of war applicable to his own service. Had Nelson chosen an army career and enlisted in one of the king's regiments at a similar age he would have enjoyed an entirely different training, educational, and experiential journey. The accumulated knowledge—coupled with his physical courage, fiery passion for battle, sharp intellect, and instinctive leadership skills—might well have enabled Nelson to emerge as one of the era's greatest soldiers. Nature certainly bestowed upon him a better package of gifts and talents than those it gave to the duke of Wellington, who nonetheless squeezed every last drop of practical ability from those he received, and deservedly earned his nation's accolades as "the most successful soldier in English history."[125] Nature's endowments to Nelson were strikingly similar to those given to the maneuverists Erwin Rommel and George S. Patton. It is no wonder, then, that as a *sea* officer Nelson stands with them among history's greatest warriors.

6

Coalition Warfare

SINCE THE rise of the modern nation-state in the seventeenth century, an enduring feature of European warfare has been its multinational character. Nation-states have seldom fought in major wars by themselves, but they have almost always sought to form coalitions or alliances of willing partners with similar strategic outlooks. One notes, for example, that the major wars of the twentieth century involved clashes between opposing coalitions (World Wars I and II, the Korean War, the Vietnam War); standoffs between opposing alliances (the Cold War); campaigns between a single nation and a coalition (the Six Day War, the Yom Kippur War, the 1991 Persian Gulf War); or between a single nation and an alliance (the North Atlantic Treaty Organization's [NATO's] 1999 "Kosovo War").

It was the same in the preceding centuries. The major European wars of the eighteenth and nineteenth centuries (the Seven Years' War, the French Revolutionary Wars, the Napoleonic Wars, and the Crimean War) involved clashes of coalitions. A coalition, according to the definition used in this book, is an arrangement between two or more nations to achieve specific, usually short-term foreign policy goals, normally formed hastily in response to real or threatened aggression by another nation.[1] Without the explicit agreements, rules, and protocols that underpin a formal alliance, a coalition can disappear virtually overnight if one or more of the partners has had enough. While friendship between partner nations is desirable, and will enhance the cohesion of the coalition, a state of friendship is by no means assured or required.

The definition just provided describes perfectly the seven multinational military coalitions that waged war against revolutionary and Napoleonic France between 1792 and 1815. Although seven distinct coalitions rose and dissolved, not all participating nations treated them with the same degree of seriousness. Some of the smaller or less politically stable participants entertained a highly expedient view of the coalitions, and left and reentered them several times (as Spain did) in order to stay on what appeared at any given point to be the winning side. Great Britain was the only nation to belong to all seven anti-Gallic coalitions.

This book has already identified several periods, particularly during the First Coalition, which survived from April 1792 until October 1797, when Nelson often privately and sometimes openly cursed Britain's allies for their passivity, inadequacy, lack of aggression, and in-house squabbling. Perhaps he did not realize that the coalition partners agreed to work together for short periods only because they faced immediate threats and wanted strength through numbers. Some of the partners did not actually like each other, most had different political ambitions, and few would consent to their forces being commanded by generals or admirals from other nations.

Nelson became so frustrated by what he considered the inactivity and petty jealousies of Britain's allies that, during occasional moments of depression or frustration, he even asserted Britain should withdraw from the First Coalition and make a separate peace with the French, who could continue killing themselves for all he cared, and leave the allies to fend for themselves.

Throughout the period of the First Coalition, however, Nelson was a relatively unknown and junior sea officer, without the status or opportunities to influence relations between coalition partners. As a captain or commodore, he was one of very many Royal Navy officers who were experienced enough to see and understand regional relations but still too junior to make a significant impact, positive or negative, on the way allies cooperated or fought together.

This chapter analyzes Nelson's role in the functioning, particularly at sea, of the Second Coalition, which began in late 1798 and ended with the French-Austrian peace of February 1801. For the first half or more of this period, Nelson served as a powerful coalition fleet commander and for several months as British commander in chief in the Mediterranean. He found himself commanding British, Neapolitan, and Portuguese squadrons while trying both to frame wider strategy with Russian and Turkish naval forces and to undertake combined operations with all or some of them.

The chapter attempts to determine whether Nelson, an admiral of influence and a national hero by the time of the Second Coalition, recognized the importance of building strong relations and establishing and maintaining goodwill in multinational naval coalition contexts. In light of the fact that Nelson was exceedingly patriotic and often critical of the skills and efforts of Britain's allies, the following also analyzes his ability and willingness to see other national viewpoints, adopt conciliatory policies, set common priorities, and establish multinational strategies when necessary.

From the Nile to Palermo

The period immediately following the Battle of the Nile on 1 August 1798 brought Nelson to the forefront of Mediterranean politics for the first time. The period began well enough; after two weeks on the Egyptian coast, Nelson decided he would return to the Kingdom of the Two Sicilies for a week or so (which became two years).[2] He needed to have his flagship, the *Vanguard,* repaired in Naples and in the company of the Hamiltons wanted to recover from exhaustion, combat stress, and a head wound.

Nelson divided his victorious squadron in order to meet several objectives. He sent Captain Saumarez with seven ships of the line to Gibraltar as escort for five French prizes taken at the Nile.[3] St. Vincent, still commander in chief of the Mediterranean Fleet, even though he remained off Cadiz or at Gibraltar, would decide what to do with the prizes. Nelson gave then-Capt. Samuel Hood three other ships of the line, three frigates, and instructions to prevent Napoleon's Armée d'Orient from receiving reinforcements or supplies.[4] It must starve in the desert, Nelson said. Nelson himself took his three most damaged ships, including the *Vanguard,* to Naples for repair.

Even before he reached Naples on 22 September 1798, Nelson began organizing the security of the entire Mediterranean naval theater, worrying most about Egypt, the Ionian Islands, and Malta (all still in French hands). He also turned his attention to the urgent provisioning of his worn-out squadron. He therefore sent a series of written requests for military, naval, and logistical assistance to regional civil and military leaders as far apart as Naples, Russia, Portugal, and the Ottoman Empire (usually known simply as Turkey). Carried by sloops, corvettes, and other fast ships, these dispatches and letters often took weeks or more to reach their recipients. Waiting for specific replies sometimes required the combat-scarred hero to remain patient for a month or so.

On 8 September 1798, Nelson sent a message to the young rear admiral Don Domingos Xavier de Lima, the marquis de Niza. The thirty-three-year-old marquis, Nelson's junior by seven years, commanded a small Portuguese naval squadron that had recently left Naples for the eastern Mediterranean. Nelson asked Niza to take over blockade duties in Egyptian waters from Captain Hood. Nelson doubtless already knew from Sir William Hamilton that Niza had professed a willingness to join forces with Nelson as soon as his whereabouts became clear.[5]

In his 8 September letter, Nelson explained to Niza that Hood's small force was incapable of staying outside Alexandria for much longer. So it would be most helpful if the marquis's force of four ships of the line (all with seventy-four guns) and various other vessels would take that force's place. If Niza's squadron remained on patrol outside Alexandria for six weeks at the most, preventing anything getting in and out to support Napoleon's army, Nelson would promptly return Hood's replenished vessels to relieve Niza.[6] He could then move his squadron elsewhere as he wished.

To foster good relations and enable clear communication between the two nations' forces, Nelson, in accordance with St. Vincent's earlier instructions, dispatched to Niza Capt. Richard Retalick, a congenial multilingual officer. (Niza's written English was very poor, and his several English-speaking commodores and officers were not always conveniently placed to draft or read his letters for him.)[7] Retalick and two signal lieutenants were to act as a small liaison staff.[8] Retalick, although an Englishman, had to remain subordinate to Niza and never challenge his views unless specifically asked for advice.

Many of the letters and dispatches that Nelson would receive from Niza and other Mediterranean leaders nonetheless came only in French (then the language of European diplomacy) or Italian. Nelson, a poor linguist, thus had to have his secretaries or the multilingual Emma Hamilton translate them into English for his comprehension. Later, in the years following the Battle of Copenhagen, he routinely used his friend and chaplain, Dr. Scott, who was an exceptional linguist, to act as his interpreter and translator. Scott later recalled Nelson's sensitive handling of language differences between coalition partners and other nations. He "made it a point of etiquette to accompany all his original English letters to foreign courts with translations in their respective languages."[9]

Nelson certainly did recognize the need for courteous and sensitive communication between allies. He had noted with displeasure back in 1793 how a particular commodore had "amazingly neglected those etiquettes which at all times I should consider necessary" in foreign ports or when dealing with allies.[10] He thereafter went to great lengths to ensure he made no similar mistakes. He had foreign language newspapers translated for him, so he could remain abreast of the desires, needs, and interests of his allies; and he often purchased—usually from his own pocket—carefully chosen gifts for his allies, which actually reveal a surprising sensitivity to their cultural and religious beliefs.

Still, because he presumed Niza's Portuguese squadron to be second rate[11] (as did Sir William Hamilton, who had witnessed both Niza's lack of expertise and the high desertion rate from his ships),[12] Nelson could not resist mildly patronizing the Portuguese aristocrat in a letter of 8 September 1798. What a great pity, Nelson smugly wrote, that the marquis had not chosen to combine forces with him *before* his glorious victory at the Nile. But, he added, "as that is past remedy," and Niza surely regretted it, he could now serve "the common cause" by cutting off Alexandria from outside assistance. This action would ensure that Napoleon could get no supplies.

Nelson informed the marquis that he had also written to the grand signior (Selim III, the sultan of Turkey) in Constantinople, asking him to send the Ottoman army into Syria and the fleet into Alexandria in order to destroy Napoleon's isolated army. "That your excellency may be a partaker in these joyful events," Nelson wrote more politely on 8 September in closing to Niza, "is the sincere wish of your most obedient and faithful, humble servant, Horatio Nelson."[13]

Even before he sent this letter, which probably took a full week by swift vessel to reach its recipient, Nelson had advised Hamilton that he doubted Niza would "obey" him. "However," he added, "I will do my duty in representing the importance of the business, and if our allies will not assist in completing what has been so gloriously begun, it is not my fault, and too late they will repent it."[14] These uncharitable comments (certainly not intended for a wider readership) illustrate Nelson's initial distrust of Portuguese resolve. More importantly, they reveal his belief: as commander of the largest naval force in the central and eastern Mediterranean, he had a right to expect obedience—or "compliance" as he described it to St. Vincent—from smaller allied squadrons.[15] This expectation even applied when the allied commander, in this case Niza, was of the same rank. Both were rear admirals.

This concept that the nation providing the preponderance of forces and resources in-theater should provide the commander of the coalition force as well is now known as Lead Nation Command.[16] Nelson never knew it in these terms, of course. He merely considered it common sense, as did St. Vincent and the Portuguese government itself,[17] that he should assume overall command responsibilities since his own force was the biggest in-theater. He doubtless also felt that he had a moral right to do so after having just thrashed the French at the Nile with no multinational contributions. This attitude certainly helps explain his smug comment about the battle to Niza.

These ideas on command of a coalition logically grew out of a broader and more significant coalition concept, now known as *proportionality of power.*[18] This concept was recognized well before Nelson's time (it existed during the Seven Years' War, for example), even though it was not articulated in today's clear fashion. The concept suggests that the larger the force a nation contributes to a coalition or alliance, the greater say it should have at the strategic level in formulating the general planning and command of campaigns and in deciding the roles and responsibilities of all multinational assets. Proportionality has long been a source of discontent in coalitions and alliances. Smaller partners often feel disrespected, marginalized, excluded, or even bullied by the partner that considers itself the leader.

In September 1798, however, no new coalition yet existed. So Nelson's letters to Russian, Turkish, Neapolitan, and Portuguese leaders were, despite his talk of "obedience," no more than polite requests from an acclaimed but junior admiral. They certainly lacked the authority of, say, instructions given in a formal multinational command chain established by diplomatic agreements. This was nonetheless a period of rapid political change, and the Second

Coalition was not far in the future. Its emergence was actually an inevitable consequence of Nelson's victory at the Nile, the excitement it caused, and the appeals for assistance he wrote.

By way of brief background, the signing of the Treaty of Campo Formio in October 1797 ended five years of war between Austria and France. The First Coalition dissolved, leaving Britain (and an inert Portugal) alone at war with France. Still ruled by the Directory (not yet by Napoleon), France was doing well. The Treaty of Campo Formio gave France the Ionian Islands (off the Ottoman Turkish Adriatic coast) as well as Venetian Albania.[19] In February 1798 France occupied the Papal States in Italy and in April took control of Switzerland. These acts were not sufficient in themselves to tempt an exhausted Austria into resuming hostilities, but they did convince the new tsar of Russia, Paul I, that France sought to dominate the Mediterranean world in addition to western Europe.

Primarily aimed at disrupting British interests, Napoleon's June 1798 invasion of Malta and the following expedition toward the Levant nonetheless angered the Ottoman Empire. The Turks had recently watched with alarm as France swallowed up the Ionian Islands, just off their coast. They took Napoleon's eastward move as a clear sign of aggressive rapacity. So did the Russians, who were equally angry. Tsar Paul mobilized over ten thousand troops on Russia's southwestern frontiers and began seeking an alliance with Turkey.[20] Paul sent a naval squadron to Constantinople, commanded by Vice Adm. Fedor Fedorovitch Ushakov, a distinguished fifty-four-year-old seaman. The squadron's main job was to keep Napoleon's fleet out of the Black Sea should it attempt an anti-Russian or anti-Turkish attack.[21]

News of Napoleon's landing at Alexandria, and then of Nelson's great victory at Aboukir (which reached Constantinople on 12 August), spurred the Turkish government into declaring war on France.[22] It also prompted Ushakov and Vice Adm. Abdul Kadir Bey, the sixty-year-old commander of the Turkish squadron, to begin rapid preparations for a naval response. They would join forces and lead their combined fleet under Ushakov's overall command through the Dardanelles into the Mediterranean. They would then launch a Russo-Turkish attack against French forces occupying the Ionian Islands. Most major European powers wanted control of those islands, or at least their removal from French control.[23] Guarding the entrance to the Adriatic, the islands were strategically important for naval protection of the eastern coastlines of the Italian states and the western coastlines of Turkey.

The Russians and Turks certainly had the means to seize the Ionian Islands. Their fleet—which on paper was easily as powerful as Nelson's—included twelve ships of the line, at least seven frigates, and a miscellany of other vessels; they also had a landing force of several thousand men. To facilitate clear communications, Ushakov sent Kadir Bey a small liaison and signals team fluent in Turkish.[24] The powerful combined Russo-Turkish force passed through the straits on 1 October 1798 and began blockading the Ionian Islands four days later.[25]

The remarkable détente between Russia and Turkey reveals the expedient nature of many coalitions. These two countries were hardly natural allies, having fought five wars against each other during the preceding century. But now their hatred of republican France and its expansionist policies had grown stronger than their historical mutual hostility. So, deciding that the benefit of joining together against France outweighed the need to maintain their own squabbles, they simply put away their unresolved differences (at least for the time being). It is no wonder that other potential allies, particularly Austria and Britain, treated the Russo-Turkish friendship with great skepticism. Nelson certainly did, along with most of his peers.

In the meantime, Nelson had changed his mind about where the marquis de Niza's small Portuguese squadron should serve.[26] The marquis had sailed to Egypt in apparent obedience to Nelson's earlier "order" but, to the Englishman's disappointment, had inexplicably refused to remain off Alexandria in place of Captain Hood's worn-out ships or even to give them water.[27] So much for Nelson's orders. Still, Nelson could not ignore the fact that the four Portuguese ships of the line and their supporting brigs and frigates would be very useful off Malta—if only he could get Niza to accept his own responsibility toward the common cause.

On 13 September Nelson therefore decided to make another written appeal for assistance, this time without the patronizing and slightly bossy style that had got him nowhere in his first communications to Niza. After the English admiral learned that three French ships (one ship of the line—*Le Guillaume Tell*—and two frigates) that had escaped from the Battle of the Nile were at Malta, Nelson politely outlined to Niza the great benefit to be gained by having his squadron hunt for those ships. Nelson explained that they should do so while blockading the approaches to Malta and its small neighboring island, Gozo. The Maltese were rebelling against their French occupiers, who still clung tenaciously to La Valetta, the island's capital. The Maltese needed all the help they could get to dislodge the French. The few British ships blockading Malta also needed help. They were worn-out and insufficient by themselves to maintain an effective blockade.

Nelson clearly worried that this appeal to Niza's sense of responsibility had fallen on deaf ears. On 25 September Nelson told Earl Spencer that he had a fervent hope, although it was only a "distant one," that Niza had sailed for Malta. And as Nelson told the earl in another letter four days later, he was therefore much relieved to learn Niza had "obeyed" the "order" and was indeed blockading Malta with his four battleships and various smaller vessels.[28]

These words reveal the complex nature of multinational command. Nelson knew he had legally vested authority, granted by the Portuguese government via St. Vincent, but that the on-the-scene reality was far from clear-cut. In the absence of direct, face-to-face conversations with Niza, where Nelson's strength of personality and force of argument might prove persuasive, he had to rely on written dispatches only. He thus had to write carefully and with the greatest sensitivity toward the marquis. What he described to Lord Spencer as a written "order" to Niza actually contained no instructions or even a direct request. It was merely a politely worded, deferential letter stating that it would be very useful to all concerned if Malta were blockaded.

Thus, Nelson had become aware of the limitations, in practical terms, of his authority over Niza and wisely avoided even the appearance that he was giving him direct chain-of-command orders. Replacing orders with gentle suggestions worked far better. Niza sailed for Malta. On 4 October Nelson wrote to thank him, reaffirming Niza's unique allied status by adding: "I hope it [the French garrison on Malta] will soon surrender to the united efforts of your Excellency and the ships under my command."[29] Nelson was thus saying that he would treat Niza as a partner, give him due respect, and not needlessly interfere with the employment and direction of his ships—although Nelson remained the theater commander.

On 14 October 1798, two weeks after grumbling to St. Vincent that Naples was "a country of fiddlers and poets, whores and scoundrels," Nelson sailed for Malta to oversee the blockade himself.[30] He wanted to acquaint himself "with the true state of matters there," he told Earl Spencer.[31] He found Capt. Alexander Ball and the marquis de Niza blockading well and harmoniously with their small squadrons, even though the French garrison in La Valetta had two weeks earlier rejected a summons to surrender signed by both Saumarez and Niza.

Nelson was pleased with Niza's blockade efforts off Malta, but although he considered the Portuguese rear admiral a very pleasant young fellow, he immediately noticed Niza's lack of experience and command abilities. Marquis or not, he was out of his depth and a dreadful administrator. His squadron, which had been at sea only for a short period (three months less than Nelson's squadron), was in an inexcusably poor state. It needed urgent refitting. A passenger aboard one of the Portuguese ships a few weeks later recalled that the officers included former British seamen, but the crews were "a strange medley of negroes, mulattoes, and people of different nations, without order, discipline, or cleanliness."[32]

Now that Nelson had the measure of the man, he felt able to move beyond merely making sensitive deferential suggestions by letter. Seeing that Niza had quickly fallen under the sway of an experienced commander's stronger personality and actually seemed relieved to have his expert direction, Nelson issued the marquis a polite but unambiguous formal order: "You are hereby required and directed to proceed with the squadron of Her Most Faithful Majesty under your command to the Bay of Naples, and on your arrival there to use all expedition in fitting them for sea, and victual them for three months . . . and when the squadron under your command are fitted, you will hold yourself in constant readiness for sea."[33]

Niza had fully committed himself to his blockade duties, however, and asked if he could remain at Malta. To Nelson's pleasant surprise, Niza offered to maintain his blockade duties from another ship. The British admiral politely refused, and sent him to Naples. But he long remembered Niza's dedication and, three months later, commented happily to St. Vincent how Niza had to be "torn" from Malta.[34]

Nelson's concern for the state of Niza's Portuguese squadron is not surprising. He had never been able to understand how nations that had excellent vessels could also have barely competent—or even incompetent—captains and crews. Back in June 1793, he was stunned to see Spanish ships, which were among the world's best and probably better than British ships, so "shockingly manned."[35] A month later in July 1793, he had mocked a Spanish fleet for having a huge sick list after a mere sixty days at sea. After two months, he observed, a British fleet was just coming into its prime and would still be fighting fit several months later.[36] Now Nelson was seeing the same thing again, this time with Niza's Portuguese squadron.

The state of Niza's squadron was not the only source of Nelson's irritation or his desire to send the Portuguese back to Naples for a while. Since he was a rear admiral under Nelson's overall command, Niza had maintained in a letter to Nelson dated 22 October, he ought to be treated by every officer and seaman of subordinate rank, whether Portuguese *or* British, as their superior officer.[37] He would never issue orders contrary to Nelson's, he added, but in Nelson's absence he should be free to issue instructions and demand obedience. He told Nelson that he felt hurt, in particular, that he had not yet received formal command of Ball's ships off Malta.[38]

Despite his best efforts to work amicably with his allies, Nelson would not yet contemplate putting British ships under Niza's command. He was not opposed to British captains serving under foreign squadron commanders.[39] Yet, he considered the Portuguese squadron well intentioned but "totally useless," and Niza, although a thoroughly decent chap, still "completely ignorant of sea affairs."[40]

Nelson had also come to realize that Niza and his squadron's officers all carried rank far more senior than their experience and responsibilities warranted. A British squadron of Niza's size and importance would have a senior captain in charge, probably flying the broad pendant of a commodore. Each ship of the line would have a captain. Niza, on the other hand, wore the rank of rear admiral (despite his limited experience and short term of service), and his ships of the line all had commodores.[41] Nelson could not, then, entrust him with the responsibilities of a *real* rear admiral, even though the British admiral would not offend the marquis and cause coalition disharmony by pointing this out.

He therefore politely told Niza that he would indeed treat him as a British subordinate, which meant that he had no authority except that given to him by Nelson himself through specific orders.[42] In the meantime, Nelson gently said, Niza would be receiving command of no British vessels. He could thus issue orders only to his Portuguese subordinates.

Nonetheless, Nelson began delegating particular tasks to Niza that he thought the marquis and his commanders could manage and that might, he hoped, steadily raise their level of competence to something near British standards. That was Nelson's aim: to improve the Portuguese squadron to the point where he could trust it to perform its assigned tasks alongside British ships effectively and with minimal direct supervision.

The British rear admiral also had another consideration in mind; he had learned that Admiral Ushakov's strong Russo-Turkish fleet had entered the Mediterranean. Led by two vice admirals (thus outranking Nelson) and possessing more warships than he did, the presence of that allied fleet posed significant proportionality of power questions. Nelson was especially keen, therefore, to maintain close control of Niza's little Portuguese squadron, which (combined with his British squadron) would certainly give him numerical equivalence, if not a slight superiority. Nelson's status as senior commander would be difficult to assert in terms of rank alone, given that he was only a rear admiral. Yet his status surely rested on the facts that he was a proven hero, had just won the most devastating naval victory in memory, commanded a large force now including Portuguese allies, and had been active in-theater longer than anyone else. This, at least, is how he saw the situation.

St. Vincent had already dispatched a new set of Admiralty instructions to Nelson (dated 3 October 1798) by the time the latter had learned of Ushakov's presence and purpose in the region. These instructions were far-reaching and not entirely realistic given the size of Nelson's total force. They compelled him to protect the coasts of Naples, Sicily, and the Adriatic; to cooperate with Austrian and Neapolitan armies if they chose to renew offensive operations in Italy; to keep Napoleon's army cut off in Egypt; to blockade Malta; and to cooperate with Turkish and Russian squadrons in the Mediterranean.[43] In view of new treaties about to be signed between Turkey and Russia, Nelson was, according to the Admiralty instructions, "particularly directed, in every possible situation, to give the most cordial and unlimited support and protection to his Majesty's allies." He must also "exert himself to the utmost to preserve a good intelligence between them, and most carefully to avoid the giving of any of them the smallest cause for suspicion, jealousy or offense."[44]

This undertaking would not be easy. The Russian advance into the Mediterranean gravely worried Nelson for strategic reasons.[45] He realized that the short-term coalition balance of power was changing, apparently to

Britain's disadvantage, and that he would, if not careful, lose his status as the senior and most influential admiral in-theater. He also felt that Russia's territorial ambitions, particularly toward the Ionian Islands and Malta, would distress Naples, Britain's key Mediterranean ally. If Russia were to gain those islands—and Tsar Paul seemed likely to want Malta after having recently been elected grand master of the Knights of St. John, based in La Valetta—it would probably establish itself permanently in the Mediterranean. Russia would then not only dominate Turkey, but also become a significant rival and possibly a threat to British trade and political interests.[46]

Nelson's concerns mounted when he learned in early November 1798 that Ushakov had not taken his combined fleet to Egypt to finish off Napoleon's stranded Armée d'Orient, as Nelson had been hoping.[47] Ushakov had instead sailed his fleet toward the Ionian Islands and seemed intent on seizing them for Russia.

Nelson had been planning to sail for the Ionian Islands himself, even though he had canceled his first trip there (originally scheduled to follow straight on from his mid-October voyage to Malta)[48] so that he could spend more time at the court of Naples. He had also received an appeal for assistance by one group of Ionian islanders. And he was actively trying to persuade all the islanders to accept Britain's protective stewardship.[49] Learning of Ushakov's destination thus prompted an immediate response. Nelson dispatched his trusted Captain Troubridge to the islands "without a moment's loss of time" to claim them for Britain before Ushakov could do so for Russia.[50] He suspected he was too late (he was!) so instructed Troubridge that "should the Russians have taken possession of these islands and be cruising near with the Turkish fleet, you will pay a visit to the Turkish admiral [only]."[51] With the greatest courtesy and deference, Troubridge should, he added, ascertain the situation between the Russians and the Turks and whether the Turkish admiral was of sufficient rank for Nelson or Troubridge to hold confidential talks with him.

These requests to Troubridge reveal Nelson's desire to woo the Turks, who might yet send their squadron to Egypt, possibly under Nelson's command. Indeed, he would like nothing better. Still, the last thing he wanted to do was upset or embitter the Russians; and a week later on 17 November, he wrote to both Ushakov and Kadir Bey. His letters were extremely courteous and full of expressions of good wishes and hopes for united actions. He had recently promised Earl Spencer that he would "take care to keep on the very best footing with all the Allied Powers," and he was determined that he should not be the cause of British embarrassment or coalition tensions.[52]

Yet his letter to the Turkish admiral differed in subtle ways from that written to Ushakov. To Kadir Bey he expressed great respect for the Ottoman armed forces (he said nothing similar to Ushakov about the Russian military), adding that he would feel happy if Kadir Bey would tell him how he, Nelson, could be "most useful" to him.[53] In other words, Nelson was assuming a remarkably collegial and almost subordinate posture. He was just as courteous, but nowhere near as deferential, to Ushakov, merely telling him that he felt happy they were in close proximity and "working together for the good cause of our sovereigns."[54] Nelson doubtless wrote this last phrase with careful deliberation; it expressed in a nonconfrontational manner the notion that Ushakov should do nothing unsanctioned by Tsar Paul or damaging to British interests.

Ushakov was equally pleasant in his return correspondence but refused to be budged from the Ionian Islands. French garrisons on Cerigo, Zante, Santa Maura, Cephalonia, and Ithaca had all surrendered to him and Kadir Bey. Ushakov, whose Orthodox Christian faith was the same as most islanders, found himself hailed as the liberator. Yet, to his credit, he insisted that Kadir Bey, a Muslim, receive the same honors as he did.[55] The local inhabitants reluctantly complied.

Corfu's strong French garrison of thirty-five hundred men still held out. It gained support from the seventy-four-gun *Généreux,* one of the two French ships of the line to escape from the Battle of the Nile, and from the fifty-gun *Leander,* a British ship taken as a prize by the *Généreux,* shortly after that great battle. Ushakov intensified his efforts to take Corfu, this last and most important Ionian island. By mid-November 1798, the Russian admiral had initiated a tight blockade around Corfu with eight ships of the line and seven frigates.[56] This task seemed wasteful and extravagant to Nelson, who could think of far better ways of using such a powerful force. Yet to Ushakov, who held deep fears that the French in Italy would attempt to relieve Corfu, it was essential.[57]

Nelson was unusually patient with Ushakov, at least for a few weeks—mainly because of distractions. After his squadron at Malta had forced the French garrison on Gozo to capitulate on 30 October 1798, the Englishman had become very preoccupied with the organization of naval support for a new allied joint offensive in northwestern Italy. That offensive, scheduled for late in November, aimed at capturing Leghorn and other towns around the Gulf of Genoa. Its purpose was to cut off French lines of communication and withdrawal while a fifty-thousand-strong Neapolitan army under Austrian general Mack moved northward from Naples toward Rome. To provide naval support for this offensive, Nelson relied heavily on the Portuguese *and* Neapolitan squadrons, over which he now exercised authority through direct but always courteous commands.[58]

King Ferdinand had recently paid a fortune to expand and greatly upgrade the Neapolitan naval force, but its crews and commanders lacked the training, discipline, experience, and esprit de corps of their British counterparts. Ferdinand was thus delighted to be able to place his small fleet under the expert command of his beloved Nelson, who was presumably pleased to increase his combined fleet's numerical advantage over Ushakov's. This delight was not shared by everyone in Naples. Commo. Prince Francesco Caracciolo, the Neapolitan naval commander, rightly considered himself an experienced, professional, and effective officer. He did not like being placed under a foreign admiral, even the great Nelson, whose influence over the Neapolitan's royal family he began to resent. Yet, despite having to swallow his pride, and still nursing a bit of a grudge against Nelson for unwittingly denying him an opportunity for glory in an engagement over five years earlier, Caracciolo apparently never openly refused to obey the British admiral's orders.[59]

The joint campaign late in November 1798 went well, with Nelson's multinational coalition fleet and its transports carrying over five thousand troops north. Faced by them, the French at Leghorn quickly accepted a summons to surrender. Meanwhile, the Neapolitan army of fifty thousand under Mack was making good progress further south, liberating Rome from the French late in November. This same Neapolitan army would soon be driven back, humiliatingly, by a French army one-fifth its size. Yet in the meantime, allied progress seemed very pleasing.

With the French apparently suffering serious reverses in Italy but remaining largely intact in Egypt, Nelson on 12 December 1798 finally asked Ushakov directly to shift his focus away from Corfu and onto the Egyptian coast.[60] After outlining the recent successes of Britain and Naples—so that Ushakov would be in no doubt that Malta was off-limits—Nelson mildly rebuked him for sending only two frigates and ten gunboats to Egypt. Captain Hood's blockading vessels were in desperate need of replacement. Ushakov should have sent at least three ships of the line, four frigates, and the gunboats. "Egypt is the first object," Nelson emphasized, "Corfu the second."

Ushakov had his hands full at Corfu and politely but firmly said so. His siege of Corfu's strong fortress was not going well, his ships were being battered by a dreadful winter, and his crews were suffering a very high sick rate. That Ushakov was able to maintain his blockade throughout the winter, without the benefit of sheltered harbors (as half of Nelson's ships had), was itself quite an accomplishment.

In his letters during this period, Nelson did not address Ushakov as commander in chief of a combined fleet but merely as commander of the Russian squadron. In this way the British admiral may have been signaling that he could still legitimately assert senior coalition commander status: he was himself the legally vested commander of a combined British, Portuguese, and Neapolitan fleet with more in-theater ships than Ushakov.

Nelson did not trust Ushakov for a minute and remained convinced that the Turks, for whom he had a genuine warmth, were being exploited by the Russians. As he confessed to John Spencer Smith, the British minister at Constantinople: "I have had a long and friendly conference with Kelim Effendi [the Turkish foreign minister] on the conduct likely to be pursued by the Russian Court towards the unsuspicious (I fear) and upright Turk."[61] The Turkish government, he thought, ought to be aware of the grave danger of allowing the Russians to get a footing at Corfu.[62] He likewise told St. Vincent that the "good Turks" would probably have relieved Hood off Egypt but that "the Russians seem to me to be more intent on taking ports in the Mediterranean than destroying Buonaparte in Egypt."[63]

Nelson's affection for the Turks was sincere, and he certainly shared the sultan's view that the French were "the Enemies of God; men without Faith or Law."[64] He wrote repeatedly to Turkish leaders and statesmen, with warm expressions of respect and friendship. These letters and others written to Islamic leaders elsewhere in the Mediterranean are superbly crafted examples of diplomatic prose. They are also marvelous expressions of religious tolerance. Devoutly religious himself, Nelson seemed to appreciate the faith and devotion exhibited by all "true Musselmen" (that is, Muslims) that he worked alongside or with whom he corresponded.[65] He felt almost as determined to protect the "true faith of Mahomet" against the ungodly French (the "atheist assassins and robbers," as he eloquently described them) as he did to protect his own Christian faith from the same foes.[66]

Actually, he seldom mentioned either Christ or Muhammad (thus keeping religious differences absent), but referred to God the creator—meaning the same God who watched over the lives of both Christians and Muslims—in the most adoring manner. To the sultan, Nelson promised he would always pray "to the God of Heaven and Earth to pour down his choicest blessings on the Imperial head, to bless his Arms with success against all his Enemies, [and] to grant health and long life to the Grand Signior."[67] To the grand vizier, who effectively ran the Turkish government for the sultan, Nelson wrote:

> When I first saw the French Fleet . . . I prayed that, if our cause was just, I might be the happy instrument of His [God's] punishment against unbelievers of the Supreme only True God—that if it was unjust, I might be killed. The Almighty took the Battle into His own hand, and with His power marked the Victory as the most astonishing that ever was gained at sea; All glory be to God! Amen! Amen! . . .
>
> That your Excellency may long live in health to carry, by your wise councils, the glory of the Ottoman Empire to the highest pitch of grandeur, is the sincere prayer of your Excellency's most faithful servant, NELSON.[68]

Likewise, to Admiral Bey he wrote: "The Glory of the Ottoman Arms is as dear to me as those of my own Country, and I always pray to the God of Heaven and Earth for His blessing on the Grand Signior, and for all his faithful Subjects."[69]

Many readers might today feel that Nelson was being insincere, merely flattering the Turks and issuing empty platitudes about the Islamic faith. Yet his letters so strongly resemble those written to his loved ones about the role of God in his own life and his nation's fate that we should not see them as anything but sincere. He continued writing to the sultan, the grand vizier, and other Islamic leaders long after he had any use of direct assistance, and his letters were just as warm toward their faith.

Even if this view is wrong, and Nelson was merely paying courteous lip service to the Turks and their religion, it should not be seen as a negative thing. Cultural awareness and sensitivity remain key ingredients in successful coalition building as, for example, Gen. H. Norman Schwarzkopf continually had to remind himself during the 1991 Persian Gulf War. And when compared to Schwarzkopf, or some other twentieth-century multicultural coalition force commanders, such as Lord Louis Mountbatten and General of the Army Douglas MacArthur, Nelson actually appears more sensitive to the cultural and spiritual attitudes of his coalition partners. It bore fruit; Nelson became a favorite of Mediterranean Islamic leaders, who seemed more willing to assist him than other Western European warriors.

The Evacuation of the Neapolitan Court

Despite the rapidly deteriorating situation within Italy during mid-December 1798, with French forces evicting the Neapolitan army from Rome and driving all before them as they swept toward Naples, Nelson remained convinced that the Levant was vitally important. Napoleon's army must be destroyed. He told Kadir Bey in a letter written 17 December 1798 that he hoped that part of the Russo-Turkish fleet had gone to Egypt to help finish off that army before it could ever become a threat again. This charge was "the first object of the Ottoman Arms; Corfu is a secondary consideration."[70] Unlike his letter to Ushakov that contains an almost identical statement, this one to Kadir Bey shows not the slightest hint of accusation or reproach.

He appealed both to Kadir Bey and (via John Spencer Smith) to the Ottoman government for a Turkish squadron to sail for the Levant. It was urgently needed to help Sir Sydney Smith, John's brother and a British captain who had just arrived there. Nelson had given up asking the Russians; they were having difficulty taking Corfu anyway.

During the weeks before and after Christmas 1798, Nelson was himself fully occupied, and not happily. French forces had smashed south toward Naples, forcing Nelson (on 23 December 1798) to evacuate the royal household and the Hamiltons to Palermo, on northwestern Sicily. Nelson was distraught about the Neapolitan reversal of fortune—"My spirits have received such a shock that I think they cannot recover," he wrote[71]—and complained to anyone who would listen that King Ferdinand's nobles and aristocrats were unfaithful traitors who deserved nothing better than to be ruled by the French.[72] Yet he knew that French successes in Italy could not go unchallenged and that he would sooner or later have to attempt or at least assist with the reclamation of Ferdinand's Italian kingdom. In the meantime, he sulked in the company of the Hamiltons and commanded his hard-pressed naval force from the Hamiltons' Sicilian residence, the Palazzo Palagonia, instead of from his flagship, as his subordinates might have expected.

Interestingly, throughout this period of crisis and disappointment, Nelson had come to rely heavily on the marquis de Niza, whose small Portuguese squadron had recently proven dependable and generally useful as it transported troops to Leghorn.[73] When he had first realized that Naples was about to be overrun, Nelson immediately began planning for the evacuation of the royal family, much of its treasure, and many key persons. These, he decided, would travel to Palermo on British vessels, rather than on their own less reliable and possibly mutinous ships. Although the Neapolitan naval commander, Francesco Caracciolo, was loyal and trustworthy, his crews might not be.[74] Nelson thus gave Niza orders for the destruction of almost all the Neapolitan vessels, except for Caracciolo's flagship and one or two others in seaworthy condition. None, he said, could fall intact into French hands. As a result of what Caracciolo believed to be his king's (or possibly only Nelson's) distrust of his crews and the imminent destruction of his beloved ships, Caracciolo sank into deep depression. One observer commented that she "never saw any man look so utterly miserable."[75] This depression led to a chain of events that would soon prompt Caracciolo to change sides and eventually forfeit his life, hung by Nelson as a traitor in July 1799 from the yardarm of the *Minerva*, one of Nelson's frigates.

Nelson's failure to see the disastrous impact of his decisions on Caracciolo's morale represents a serious oversight. As even young cadets now learn, morale is a critical function of leadership. It flows downward in the chain of command. Nelson should have foreseen that Caracciolo would consider it a serious professional insult and a personal loss of face that his own king was not sailing in one of the Neapolitan ships, and that the fate of those ships was to be placed in the hands of a Portuguese officer. If Nelson ever valued Caracciolo, as Ferdinand and Maria Carolina long had, he certainly should have sought a way to achieve the desired result—getting the royal family safely to Palermo—without wounding the dignity of a fellow officer and respected ally.

Yet, Nelson was always concerned about the morale of British officers. During the Trafalgar campaign of 1805, for instance, he even let Sir Robert

Calder return to England to face a court martial in his ninety-gun three-decker—with which Nelson could hardly afford to part now that battle loomed—because he worried so much about Calder's morale. But it appears that he seldom felt the same concern for allied officers. At least he did not for Caracciolo, toward whom he displayed inadequate sympathy and sensitivity.

He was perceptive enough, however, not to ask Caracciolo to burn his own ships; either that, or he didn't trust him to carry it out. In any event, Nelson gave the unpleasant job to Niza. He may have considered Niza an inexperienced admiral, but he liked him personally and had come to value him as a trustworthy subordinate (even if Nelson still considered Niza's crews somewhat inferior). A visitor to the Neapolitan court at this time remembered Niza as a "well-bred, good-natured man, much liked by all his acquaintances." He was, more importantly, "on the best terms with Lord Nelson and all the officers of our fleet."[76]

Nelson's instructions to the Portuguese admiral were unusually firm, revealing the importance Nelson attached to the mission. Niza must order his British-born commodores, James Stone and Donald Campbell, to make all preparations for the destruction of the Neapolitan vessels in the Bay of Naples.[77] These included three ships of the line *(Guiscardo, St. Joachim, Tancredi),* as well as a frigate and some corvettes that Capt. Charles Hope, commander of a British ship that had just arrived from Egypt, would destroy. Under no circumstances, Nelson emphasized, was Niza to sail from Naples until these vessels had been destroyed. Only then could he sail as quickly as possible to Palermo, where he would receive further instructions directly from Nelson.

One cannot doubt the importance that Nelson attached to this mission. After all, he had placed Niza in charge of Captain Hope and his crew.[78] This was quite a concession, allowing the young Portuguese admiral formal command of British seamen for the first time. Yet two days later, Nelson changed his mind, or had his mind changed, about the fate of Caracciolo's warships. Apparently, after instructions from King Ferdinand, he informed Niza in writing that the Neapolitan warships should merely be moved farther out in the harbor, beyond the Portuguese ships. Because the Neapolitan crews were already deserting and could not be trusted anyway, Nelson told Niza to man the Neapolitan warships with British and Portuguese skeleton crews. Then, fitted with jury masts if necessary, the ships should be sailed to the safety of Messina, on the northeastern tip of Sicily, as soon as the ships were able.[79] They must not be destroyed unless their capture seemed probable.

This was a bitterly frustrating time for Nelson, who felt that the Neapolitan catastrophe had rubbed some of the polish off his great victory at the Nile. It got worse. When Earl Spencer had decided to send Sir Sydney Smith to help in the Levant, he failed to make clear to the ambitious, self-centered fellow that he was to come under Nelson's overall authority, even though Smith would assume local command of subordinate vessels off Alexandria. But Smith disregarded Spencer's directive and aroused his commanding admiral's ire. The admiral was mortified by Smith's letter to Hamilton telling him that Smith was effectively detaching Captain Hood and his small squadron from Nelson's command. Nelson disliked the newly arrived captain for his opportunistic presumption of the admiral's regional command authority (but it was indicative of Nelson's professionalism that he attempted to get Turkish ships to help a commander he didn't like). On 31 December, Nelson sent an impassioned plea to St. Vincent, pointing out the injustice of Smith's action and asking St. Vincent for permission to retire from the Royal Navy.[80] After close to three weeks, during which the rear admiral raged about Smith's wicked presumptions and sent the trusted Troubridge to assert Nelson's superior rank over Sir Sydney,[81] St. Vincent's reply arrived. To Nelson's great relief, it contained empathetic reassurances and, most important, orders for Nelson immediately to take Smith under his command.[82]

On 1 January 1799, Nelson wrote to Earl Spencer, politely pointing out that he could not afford to part with any vessels, even to another Englishman such as Smith. Although his fleet seemed large, he explained, "I always desire it to be understood, that I count the Portuguese as nothing but trouble."[83] This comment is consistent with one he made on 7 January to then-Commo. John Duckworth, who commanded a separate British squadron in the western Mediterranean. Nelson promised Duckworth two ships of the line for blockade duties off Toulon, then added: "I have under my command four Portuguese ships of the line; you are most heartily welcome to them all, if you think they will be useful." They were worthless, he exaggerated, ever causing him to waste valuable time writing orders, "which are intolerably executed" anyway.[84]

These criticisms should not be considered representative of his usual views of Niza's squadron; Nelson's views, which fluctuated between irritation and satisfaction, had, in general, steadily improved throughout the period he and Niza served together. Nelson actually made these criticisms mainly to prevent other British leaders, who might assume that he had a vast force at his disposal, from demanding or even asking that he give up some of his vessels.[85] If he could convince them that his British squadron was not that big, and that his Portuguese squadron was not that good, he might just retain the whole lot. Importantly, he made these criticisms in private, and, as a good multination commander, certainly never gave the Portuguese any reason to doubt the British commander's good will.

It may also simply be that Nelson was going through a rough patch with some of the Portuguese when he make these comments. Rumors had begun arriving in Palermo that they had grossly failed in their duty. Contrary to Nelson's final instructions on the matter, he heard that the Portuguese had burned all of King Ferdinand's fleet even though French forces had not yet entered Naples. He received confirmation that this was no mere rumor on 11 January 1799, just as the Portuguese squadron was arriving in Palermo. To his horror, he discovered that the squadron had not come from Messina, where it should have been with the Neapolitan ships, but from Naples, where it had destroyed them.

Nelson was aghast and, after receiving complaints from King Ferdinand, requested an explanation from his friend Niza. Had Niza given Commo. Donald Campbell any orders for this "positive disobedience" to the final orders given by himself? (To this Niza, who was equally upset, replied truthfully that he had not.) "I only beg," Nelson angrily wrote, "that I may not see Commodore Campbell till this very serious matter is explained to my satisfaction."[86]

When Campbell arrived in Naples on 13 January 1799, Nelson demanded to know why he had burned the Neapolitan warships. Campbell protested that he had come to distrust Gen. Frencesco Prince Pignatelli, whom King Ferdinand had placed in charge of Naples when he fled, and sincerely believed Pignatelli would surrender the ships to the French. Campbell therefore decided the best thing he could do was burn the ships while he still had the chance. His explanation did not persuade Nelson, who had yet to learn that the weaselly Pignatelli had already negotiated an armistice without Ferdinand's permission and on very unfavorable terms. Nelson severely chastised the commodore by letter, and told Niza that he anticipated him treating Campbell according to Portuguese naval custom for such circumstances.[87] Nelson doubtless hoped this meant a formal censure or reprimand of some kind, or possibly even a court-martial. Were Campbell an Englishman (that is to say, serving under his direct command), Nelson told Sir John Acton, he would order the court-martial himself. He was only sorry that he could not do so to a member of a non-British coalition squadron.[88]

Considering his authority, Nelson probably could have organized a court-martial for Campbell: Nelson commanded the Portuguese squadron with full permission and authority of the Portuguese government; and further, Campbell was technically still a British subject (even if now a Portuguese officer). That Nelson chose not to initiate proceedings, or even to demand that Niza do so or send Campbell in disgrace to Portugal, reflects well on the commander in chief. He considered coalition harmony and his relationship with Niza worth more than the need to punish this offense. Niza, meanwhile, initiated court-martial proceedings of his own before Nelson advised him on 28 February that the queen of Naples had come to accept Campbell's version of events. So had Nelson himself.[89] The matter was over, he said, and the commodore could return to duties "without any thought of what is past."

The whole affair may have reinforced Nelson's low estimation of some of Niza's subordinates, but in order to maintain coalition harmony, he never let this show. His communications with Campbell even remained cordial throughout the rest of the year, which is consistent with his Christian view that a man's slate was wiped clean should he be forgiven for an error or bad deed. Moreover, the affair did not damage Nelson's relationship with the Portuguese commander. On 25 January he discreetly wrote to St. Vincent that while the "great commodores" (Campbell and the other Portuguese captains) were undeserving of their rank, Niza was "a good tempered man" who was worthy of better subordinates than the [swear word, blanked out] he had.[90] "We are apparently the very best of friends," Nelson wrote, "nor have I, or will I, do an unkind thing by him."

Nelson kept his word and retained the very best relationship with Niza and thus, by extension, enhanced the relationship between the governments of Britain and Portugal. In February 1799, after French forces had swallowed up all of southern Italy, Nelson recognized that only the narrow Straits of Messina separated those forces from Sicily. But should the French attempt a crossing, he also believed that he alone could prevent its success, particularly as there were no reliable land forces on hand.[91] Nelson therefore entrusted Niza and his squadron with the vital mission of patrolling the straits, destroying any boats and ships on the Italian shore, and manning the fortress of Messina itself.[92] Even though Nelson then tried to raise other forces to support Niza, including the Russians and the Turks, he knew that Niza would not fail to meet Nelson's expectations.

Later, when Niza received orders in August 1799 to return his flag captain, Don Rodrigo de Pinto, to Lisbon, Nelson worried that a rumor claiming he and Pinto did not get along had been the cause of the recall. He immediately wrote Niza to affirm his respect for the flag captain and to ask Niza to decline the recall request. He also wrote to Don Rodrigo de Souza Coutinho, the Portuguese minister of marine in Lisbon. Nelson assured the minister that not only did he respect Pinto and did not want him removed, but that he had come to adore Niza, whom he had just appointed commander of all British and Portuguese ships off Malta—a tangible sign of his trust.[93] Niza's conduct had always been so exemplary, Nelson explained, "as to entitle him to the appellation of an excellent officer, and in private life I love him with the affection of a brother."[94]

This was no insincere statement. The emotional Nelson often professed love for his friends. During this period, for example, he wrote to Captain Ball: "I love, honour and respect you, and no persons ever have, nor could they if they were so disposed, lessen you in my esteem, both as a public officer and a private man."[95] It is certainly significant, though, that Nelson had come to see Niza, alone among foreign seamen, in the same light.

Even before the eventual allied reclamation of Naples and the suppression of the Jacobins in June, in which Nelson played key roles, he had switched his main focus (albeit with some reluctance) back to the protection of the western Mediterranean. Vice Adm. Eustache Bruix's strong French fleet had escaped a British blockade of Brest on 25 April 1799 and entered the Mediterranean. Adm. Don Josef de Mazaredo's Spanish fleet meanwhile also entered that great sea, causing Nelson much anxiety, especially as he felt that the Neapolitan situation was just as precarious. By late May, however, instead of uniting to defeat the British somewhere, both enemy fleets had succumbed to bad weather and fear of the British. They therefore retired in poor condition, the French to Toulon and the Spanish to Cartagena (on the Spanish Mediterranean coast, two hundred miles east of Gibraltar). British squadrons immediately threw tight blockading screens around them and breathed a sigh of relief.

With Naples back under Ferdinand's sovereignty by late June 1799, and enemy naval threats reduced, Nelson turned his attention again to Malta. He had continually blockaded the island for almost a year, although not always strongly due to commitments elsewhere. Now, determined to win Malta for Britain and Naples and to deny it to the Russians, Nelson entrusted Niza with a combined British-Portuguese squadron containing over half of all Nelson's warships east of Naples. (Two ships of the line, still under Sir Sydney Smith, remained in the Levant.) By September, Niza had six ships of the line (three of them British), along with some smaller armed vessels and supply and communications ships.[96]

Niza's blockading squadron almost wore itself out as it maintained a tight blockade around La Valetta and landed troops for siege operations; but the French garrison continued to hold on with remarkable, unexpected tenacity. Niza worked harmoniously with Capt. Alexander Ball, who eventually became governor of Malta (except for the cursed French-held fortress, which stubbornly held out long after all other French troops in Malta had fled or surrendered). Nelson had told Ball that Niza was "a good young man, [who] will afford you every assistance."[97] Ball soon came to share this assessment, and he and Niza became fast friends.

Only one potential snag existed: Portugal and Naples were at war with the Barbary States, in Ottoman Africa, even though those states enjoyed fair relations with Britain, thanks to British bribes, and were themselves at war with France. Those states—Morocco, Algiers, Tunis, and Tripoli—all conducted piracy off their shores, which forced Britain and other nations that wanted no interruptions to their seaborne trade to make payments in return for safe passage through those waters. Because of the state of war between Portugal and the Barbary States, Nelson always had to ensure that Niza's squadron kept well clear of Tunisian waters or the cruising area of the Barbary pirates.[98] The last thing Nelson wanted was for Portuguese and Tunisian warships to clash and thereby destabilize British coalition efforts in the Mediterranean. He successfully prevented that happening through careful management of Niza's deployment areas and by writing beautifully crafted letters to Barbary leaders. In these he extolled the importance of giving all friends of the Islamic God "and His Holy Prophet," *including the Portuguese,* full freedom to defeat the French "infidels."[99] Niza was equally sensitive to the issue, and compromised on several occasions for the good will of the coalition.

Niza was also sensitive to the issue of slavery, on which Portugal and Britain differed. Niza had slaves among his crews, to which Nelson turned a blind eye until he heard from Kelim Effendi that some of those slaves were captured Barbary pirates or other Moors and were thus Islamic. Effendi was upset that the Portuguese had some of his co-religionists as slaves. He asked Nelson to secure their freedom. Nelson in turn "begged" Niza—as a personal favor to a friend and to Britain itself—to grant those Muslims freedom.[100] Wishing to be a good friend and coalition partner, the marquis immediately acceded to the request. Nelson was delighted by Niza's "handsome" gesture, and promptly gave Effendi twenty-five emancipated Islamic slaves to take back to Constantinople.[101]

Niza's constant efforts to meet Nelson's expectations and prove himself worthy of ever greater responsibility certainly earned him much credit. Under Nelson's attentive and friendly leadership and with Ball on hand to give advice, the Portuguese aristocrat continued to blossom and became a credible, effective squadron commander. Indeed, he was the only foreign sea officer that Nelson would ever treat as a member of his famed "band of brothers."

Nelson was therefore gravely upset to learn in October 1799 that the Portuguese minister of marine had recalled Niza's entire squadron, believing that it was no longer needed in the Mediterranean (or so Nelson claimed Coutinho believed). Nelson, who was acting British commander in chief of the entire Mediterranean throughout this period, immediately wrote to Niza and virtually implored him not to leave with his squadron.[102] The Portuguese government's assessment of the situation in-theater was entirely wrong, he insisted, and Niza's services were in fact "never more wanted than at this moment." Nelson even promised him more ships (which would have given Niza more power than Nelson *ever* delegated to a subordinate squadron commander). In another letter written on the same day, 3 October, Nelson emphasized that the withdrawal of the Portuguese squadron from the blockade of Malta would have "the most ruinous consequences." Niza must therefore, Nelson strictly ordered, ignore the instructions from Lisbon and not remove a single ship—indeed, not a single man—until he received further orders from the commander in chief.

The conflicting orders naturally placed a great strain on Niza. He felt caught between the instructions of his own minister, who knew little of the true situation, and those of his coalition fleet commander, whose situation assessment was undoubtedly correct but whose legal authority over him was now questionable. Nelson recognized Niza's predicament and worried he might choose to leave.[103] Three weeks later, Nelson wrote his most heartfelt plea, telling Niza: "You were the first at the blockade, and I hope [you] will be first at its [La Valetta's] surrender. Your Excellency's conduct has gained you the love and esteem of Governor Ball, all the British officers and men, and the whole Maltese people."[104] He added his own sentiments: "and give me leave to add the name of Nelson as one of your warmest admirers, as an officer and a friend."

Even before the ship carrying this letter reached Niza, Nelson received one from him advising that he must with much regret soon follow the orders of his government and withdraw his squadron from Nelson's service. Nelson responded on 27 October imploring Niza "by every tie of honour to your Court, the Ally of my gracious sovereign," not to leave immediately but to wait until Nelson could find a replacement squadron, possibly even a Russian one. If the marquis left Malta before then, he added (with a touch of emotional blackmail thrown in for good measure), the Portuguese monarch might even reprove Niza for it.[105]

Niza knew he had no choice but to return to Lisbon. Yet, out of sympathy for Nelson's plight, profound affection for him, and strong commitment to the common cause, the Portuguese rear admiral delayed his departure as long as possible. He agreed not to leave until replacements came (this finally occurred in December). Nelson was touched and relieved by this gesture, and assured Niza on 15 November, with one of his unique Nelsonisms, that he saw Niza's delayed compliance with his government as "loyal and proper disobedience." Nelson thanked Niza profusely, promising him affectionate friendship forever (a promise he kept). In the meantime, Nelson desperately cobbled together a relieving blockade force under Troubridge, who replaced Niza in mid-December.

On 18 December 1799, Nelson formally released Niza's squadron from service with the fleet, writing him two emotional personal notes in which, once again, he thanked him profusely for all the marvelous work he and his squadron had done throughout the previous sixteen months. Shortly afterward with typical generosity of spirit, Nelson wrote to the Portuguese minister of marine offering profound thanks for having loaned him Niza and his warships.[106] "When I mention my brother and friend, Niza, I must say that I never knew so indefatigable an officer," he wrote. "During the whole time I have had the happiness of having him under my command, I have never expressed a wish that Niza did not fly to execute." Nelson even extended praise to the commodores who had, at least in their first few months of ser-vice, so often frustrated and annoyed him. This doubtless reassured the Portuguese government that it had sent the right man into the Mediterranean. Even today, Lisbon's Naval Museum (Museu de Marinha) proudly exhibits artifacts from Niza's squadron as well as paintings of him with Nelson.

The Russians and Turks

While all this was going on, the Russian and Turkish squadrons under Admirals Ushakov and Kadir Bey had remained in the Adriatic. After Sir Sydney Smith arrived in the Levant, Nelson's appeals to the Russians and Turks for assistance in and around Egypt had begun to decrease. In February 1799 he asked them instead, but with an equal lack of success, to send forces to Messina. He was still most concerned by their continued blockade and siege of Corfu, which seemed wasteful and unproductive. It could not even prevent the French ship *Généreux*, with its sails painted black, from escaping to French-occupied Ancona on the moonless night of 5–6 February 1799.[107] The French garrison of Corfu, though, finally realized that no reinforcements could be expected. With wood for heating and food in short supply, Corfu agreed to Ushakov's terms of surrender on 1 March 1799.

This event gave no pleasure to Nelson, even though on 23 March he wrote to congratulate Ushakov, who had now initiated both the investment of Ancona by a detached squadron and a series of small landings to probe elsewhere on the Italian mainland.[108] Nelson had come to resent the Russians as much as he had come to like the Turks and rely on the Portuguese. On 21 January he had written to Captain Ball at Malta, telling him how annoyed he was to learn that a Russian warship had been to Malta and sent some kind of proclamation to its citizens. "I hate the Russians," he wrote to Ball, insisting if the Russian ship was acting under Ushakov's orders, the admiral was nothing but a "blackguard."[109] Malta must return to Neapolitan sovereignty or, if Ferdinand so decided, to Britain.

The battle-scarred Nelson was very concerned by the threat of Russian interference at Malta. On 4 February 1799, he told Ball to make it clear to the Maltese that their urgently needed corn supplies from Sicily would immediately cease should they dare to fly a Russian flag.[110] And if Ushakov or any other Russian admiral should attempt to raise a flag on Malta, Ball was to "convince him of the very unhandsome manner of treating the legitimate sovereign of Malta" (Ferdinand of the Kingdom of the Two Sicilies) in this way. "The Russians shall never take the lead," Nelson stated. Then, when the siege and blockade of Corfu ended on 1 March, he became increasingly determined that Ushakov would keep well away from Malta, over which—as Nelson politely but firmly told the Russian on 23 March—the flags of both Sicily and Britain were already flying.[111]

Just as an aside, Nelson's mention of corn shipments to the Maltese raises an important but seldom considered point: his ever busy squadrons were not just involved in routine naval duties around the central and eastern Mediterranean; they also kept many isolated communities sustained during this time of interrupted trade and commerce. This service was not based only on humanitarian motives, although Nelson could not bear the idea of "ordinary" people suffering privations because of war. Astutely recognizing the significance of food to community morale—and thus to political allegiance—he was also very keen that every group allied to Naples or Britain would be given or sold cheaply the maximum amount of food. For example, in April 1799 he informed Sir John Acton, Prime Minister of Naples, that certain islands in the Bay of Naples simply must remain loyal to Ferdinand. They had already changed sides once. It was thus essential, Nelson wrote, "that the very greatest care should be taken that the islanders are supplied with the greatest abundance of provisions, and at the *very* cheapest rate."[112]

If they could not afford to buy food even at a reduced rate, Ferdinand's government should give it to them for free. It was a basic principle that those whose allegiance one wanted should always "have plenty to eat."

Certainly the last thing Nelson wanted to see was an appeal by the Maltese or any other community for Russian aid. Naples and Britain must therefore build friendships by meeting the needs of the hard-pressed peoples in-theater. Suffering most, of course, were the "half-starving" Maltese, who were entirely reliant on supplies from the Portuguese and British vessels which were blockading the French at La Valetta.[113] Even in peacetime, Malta—twice the size of the Isle of Wight and with a population then of ninety thousand —had imported most of its corn, meat, wine, and firewood.[114] Nelson duly organized as many food shipments as he could manage. He even sometimes bribed the Barbary leaders in order to ensure the convoys' safe passage or subtly threatened them if they failed to give it, in the hope that the Maltese would remain loyal and not become embittered. This maneuvering took up a lot of his time, much of it spent in meetings with Acton or King Ferdinand,[115] but Nelson's efforts eventually paid off. The Maltese people received a steady trickle of food and wood and thus became decidedly pro-British.

Ushakov deserves some of the blame for Nelson's deep suspicion of Russian motives. He had repeatedly rejected Nelson's requests for assistance at Alexandria and along the coast of Syria and during February 1799 ignored Nelson's pleas to send vessels to reinforce Niza at Messina. Paradoxically, Ushakov's motive for doing so was a reciprocal distrust of British motives. He complained in private to Vasili Tomara, the Russian ambassador in Constantinople, that the British expect "us [to] catch flies while they enter the places from which they are trying to keep us away."[116] Indeed, they have "always wanted to take Corfu for themselves and wished to send us away under various pretexts, or, by splitting us up, reduce us to an incapable condition."

Aware of the need for an outward show of harmony, Ushakov was as discreet with his gripes as Nelson was. He would "willingly" help out, he told Nelson, but "various circumstances" prevented him detaching ships for missions in the Levant, including storm damage and a lack of provisions.[117] Even if true, these excuses would hardly have impressed Nelson, whose own squadrons were exhausted and battered but still faithfully on duty.

Then, after taking Corfu on 1 March 1799, Ushakov refused to hand back the former British ship *Leander*, claiming it as his own—and only—prize of war. Nelson was patient and waited until Ushakov arrived in Sicilian waters almost five months later, in August, before asking again for the ship's return. He politely explained to Ushakov that the Admiralty, after the courts of St. Petersburg and London had negotiated the return of the *Leander*, had assigned a British crew to take charge of the ship as soon as possible.[118] The Russian admiral was shocked to read Nelson's letter on 22 August and again refused to hand back the *Leander*. He complained to Vasili Tomara: "I am surprised by such a shameless demand contrary to the laws of the whole world."[119] Eventually Ushakov returned the ship but only after receiving an explicit order from Tsar Paul to do so.

That does not mean that Nelson gave up on the Russians. On the contrary, by March 1799 he had begun to realize that the Russian fleet might prove invaluable if Naples were to be reclaimed for Ferdinand. Already acting independently of its Turkish allies, the Russian fleet was using only Russian ships and marines in its activities along the eastern Italian coast. Tsar Paul had recently signed a treaty of cooperation with Ferdinand and had dispatched an army to Italy led by that greatest of generals, the invincible Field Marshal Alexander Suvorov. Paul instructed Suvorov to cooperate with the two Austrian armies that had commenced operations in southern Europe. One of those armies gained a quick success by thrashing the French at Stockach (in today's southern Germany) on 25 March. But the other suffered a resounding defeat at Magnano (northern Italy) a week later and seemed destined for total destruction until saved by Suvorov's intervention. Suvorov not only turned the tide, but successively defeated French forces at Cassano (26 April), the Trebbia (18–19 June), and Novi (15 August). Throughout this period Ushakov provided some logistical and direct naval support to Suvorov via the Adriatic.

Nelson was initially pleased to see Ushakov finally starting to serve the common good (at least as *he* understood it); he later felt grateful in May when the Russian admiral responded to the threat posed by Bruix's French fleet by concentrating his own combined fleet at Corfu, while Nelson concentrated his at Palermo. Bruix's chances of invading Italy were now minimal. But once concentrated at Corfu, Ushakov's fleet, swollen by Russian and Turkish reinforcements to twenty ships of the line, swung ineffectively at anchor, refusing to sail out again. They would not help Nelson with the blockade of La Valetta or cooperate with Keith's squadrons in the western Mediterranean when the "Bruix scare" reemerged in July.

In August Ushakov finally moved from the Adriatic, apparently compelled to do so by dispatches from St. Petersburg. He arrived in the Straits of Messina on 14 August with six ships of the line, three frigates, and four dispatch vessels, joined by a further four ships of the line, three frigates, and various other Turkish vessels under Kadir Bey. After dividing some of this vast force into separate squadrons and sending them to different points around Italy and the Gulf of Genoa, Ushakov sailed on to his historical meeting with Nelson at Palermo.

The British commander in chief was excited to learn the Russians were on their way; and when the first squadron, under Rear Admiral Kartsov, arrived in Palermo on 16 August, Nelson regretted that he could not immediately send it north to provide coastal support to the "brave and excellent" Field Marshal Suvorov.[120] Nelson lacked authority to do so, he explained to a colleague, because the Russians "are not under my orders or influence."[121] This was naturally true; neither by high-level coalition decisions nor by on-the-spot ad hoc arrangements had Nelson been able to assume command of any Russian or Turkish naval forces. His command authority had never extended beyond British, Neapolitan, and Portuguese squadrons and apparently never would. He now recognized this and just wanted to make the best of the situation.

Nelson himself yearned to help Field Marshal Suvorov and, curiously, hoped that the arrival of the main Russian naval force would free up Nelson's squadron enough for him to send British warships north to aid Suvorov's campaign. Although he did not know the seventy-year-old Russian marshal personally, he informed him with typical emotional excitement that he couldn't wait to embrace him.[122] Indeed, Nelson believed from Suvorov's illustrious reputation, matched (unlike Mack's) by marvelous victories, that he was the Russian equivalent of Lord Hood or Earl St. Vincent: old but energetic, as well as wise, experienced, aggressive, tenacious, and brave.

Ushakov, on the other hand, was few of these things. At least that's what Nelson quickly came to believe after meeting him in Palermo on 2 September 1799. As well as refusing to hand back the *Leander,* the Russian admiral seemed reluctant to send a powerful squadron north to assist Suvorov's army or to help stabilize the situation at Naples. To Nelson's distress, Ushakov had his eyes on Malta. "The Russians are anxious to get to Malta, and care for nothing else," Nelson discreetly complained to Ball, whom he urged to be extra energetic.[123] He hoped Ball would take La Valetta before the Russians arrived there. Nelson was not alone in wanting La Valetta to fall to British and Portuguese forces only, without Russian help. Earl Spencer and many others in England had the same desire. As Spencer wrote to Nelson: "though I have full confidence in you and in Captain Ball for cooperating . . . cordially with the Russians . . . it would I confess give me peculiar satisfaction to see that island reduced by our naval force without the interference of any other maritime power."[124]

Nelson's only consolation—a strange one for an ally to feel—was that Ushakov's fleet was not in the best state and even according to the Russian admiral himself would have trouble remaining at sea over winter. This circumstance reinforced Nelson's notion that Ushakov was not his equal as a sea officer, despite the numerical superiority of the combined Russo-Turkish fleet (it had twenty ships of the line; Nelson had fifteen). *Ushakov* "has a polished outside," Nelson wrote to Earl Spencer, "but the bear is close to the skin."[125]

The Turkish admiral and senior officers who arrived at Palermo nonetheless gave Nelson much pleasure. He was immediately attracted to the handsome, articulate, cultured, and religious Kadir Bey, seeing in him a kindred spirit. He was, Nelson thought, in many ways a lot like Niza: decent and guileless, with great natural potential, a lack of experience that could soon be fixed, but poor crews that probably could not be. (They were "good people, but perfectly useless.") The big difference was that Niza had willingly placed himself under Nelson's command and tutelage, whereas Kadir Bey remained under Ushakov's.

Kadir Bey's squadron was not in-theater much longer anyway. Dissatisfied and undisciplined, and perhaps drunk on Sicilian wine, some of his seamen on shore leave stirred the anger of local Sicilians. Confrontations followed. The Sicilians left fourteen killed, fifty-three wounded, and forty missing from the Turkish crews.[126] At one point several anguished Turkish officers wanted to turn their ships' guns on Palermo itself.

Ushakov and Nelson were as mortified as Kadir Bey was ashamed. The Turk sank into immediate depression and confessed to believing his career was over.[127] This reaction prompted Nelson, with commendable charity, to write an immediate and glowing testimony to the grand signior in Constantinople. Kadir Bey was with him every day, Nelson wrote, "and a better man does not live in the world, or a better officer; he is my brother."[128] He assured the grand signior that the "little disturbance" in Palermo was in no way attributable to any fault on Kadir Bey's part. This letter and Nelson's repeated expressions of admiration for Kadir Bey, sent to the admiral himself as well as his political masters, doubtless saved him from a grim fate back home in Constantinople. He nonetheless dejectedly headed there on 12 September, taking his entire squadron.

Ushakov, meanwhile, sailed for Naples, convinced by Nelson, Acton, Hamilton, King Ferdinand, and his own nation's ambassador to Naples that he could better serve the common cause by taking the Russian fleet there than to Malta. The Russian warships remained at Naples for three months, seemingly as inactive there as they had been in the Adriatic, even though their presence undoubtedly helped in the reestablishment of Neapolitan law and order.

Nelson was not surprised by the inactivity, especially as Ushakov had himself confessed, and the poor state of his ships revealed, that the Russians were unable to remain at sea throughout another winter. Yet Nelson wrote to Ushakov on several occasions, reassuring him that he still looked forward to working with him again (really for the *first* time) in combined Russian-British operations, hopefully at Malta. Nelson really did not want Ushakov at Malta, but he had received a clear message from the British government that the island's sovereignty would eventually pass to Russia, not to Sicily or Britain.

Bad weather exacerbated this bad news. As winter closed in, Nelson realized that the state of his own ships and Niza's was becoming desperate. Then, when the Portuguese government recalled Niza's squadron to Lisbon, Nelson began to despair about Malta. And Ushakov would not budge. Instead, he remained at anchor in the Bay of Naples, acting very high-and-mighty (in Nelson's privately expressed opinion) and threatening to return to Corfu with his entire force.[129] Nelson responded by virtually pleading with Ushakov, his "most faithful Brother-in-Arms," to "go together to Malta."[130] This appeal and others initially fell on deaf ears, partly because Ushakov had misread the meaning of some of Thomas Troubridge's clumsy recent actions at Naples and resigned himself to the fact (as he saw it) that the British were still trying to undermine him.

Ushakov was nonetheless a professional seaman and soon felt compelled to swallow his pride and send his ships to Malta. On 4 November 1799, he informed Nelson that he would do so.[131] The Englishman waited impatiently fretting about Malta's fate before finally giving up hope in mid-December that Russian naval assistance would arrive. On 22 November he told John Spencer Smith that Ushakov "cannot be got to move; by his carelessness the fall of Malta is not only retarded, but the island may be lost." He told Captain Berry that Ushakov was simply "lazy."

This wasn't really fair; Ushakov's force *was* actually on its way to Malta, but dramatic intervention from St. Petersburg prevented any vessels reaching the island. An imperial decree from Tsar Paul, dated 3 November 1799, ordered Ushakov to withdraw his entire fleet from the Mediterranean and return forthwith to the Black Sea.[132] No explanation was given, although the decision stemmed from Paul's sense that he was backing the wrong horse. The British and other allies, he had come to believe, were not interested in offering him genuine opportunities to share glory. Even worse, the Austrians had exploited the gains made by Suvorov in northern Italy to strengthen their own position there, at Russia's expense. Then, in September, Austria refused to send delegates to a coalition congress to be held in St. Petersburg. Early in October, Paul was grieved to learn that his army in Switzerland had suffered defeat, apparently due to Austrian inertia or duplicity.[133]

Ushakov was upset by the news that Russia was withdrawing from the Second Coalition and that he had to lead his force out of the Mediterranean. "I had hoped to take Malta," he glumly wrote, "acting . . . with Rear Admiral Nelson and Captain Ball . . . but all our actions depend on the will of the All-Highest."[134] In the hope that his tsar would change his mind, the despondent Russian admiral took his vessels to Corfu for "the repair of ships damaged by a great storm."[135] And he was still there in April 1800, ostensibly preparing his worn-out ships for the long voyage to Sebastopol, when Nelson wrote to seek his assistance once more. But Tsar Paul's dramatic political about-turn was clearly permanent, and Ushakov soon after sailed his force back through the Dardanelles.

The French garrison on Malta finally surrendered to Royal Navy and British Army besiegers in September 1800, two years after the blockade first began. Despite bitter complaints from Tsar Paul, Britain chose to keep the island under its "protection." This reinforced in the tsar's mind the deep resentments he now felt for the allies and gave him confirmation that he had been right to pull out of the Second Coalition. As for Nelson, he remained in the Kingdom of the Two Sicilies until July 1800, when he finally disentangled himself from the kingdom's affairs and began his overland trip with the Hamiltons to England. His toughest challenges as a coalition commander and multinational political-strategic advisor were over.

Conclusions

Nelson learned a lot during his two years in the Mediterranean following the Battle of the Nile. For the first time he was called upon to help shape national and regional strategies, directly implement multinational political decisions, and undertake naval coalition warfare. It was a steep learning curve, and he made several significant mistakes, especially when ensnared by the charms of Emma Hamilton and the largess of Ferdinand and Maria Carolina. Yet Nelson possessed a quality common to great commanders; he might make a particular mistake once, but seldom twice, and never three times. In other words, he learned quickly.

One of the hardest lessons to learn was that not all supposedly friendly nations, and not even all formal allies, understood what he called "the common cause" in the same terms. Although he considered it logical that allied naval squadrons should finish off what he had started at the Battle of the Nile, ideally by blockading Alexandria and attacking the French transports anchored there, his Russian and Turkish allies formed other priorities. Nelson initially tried hard to get them to see the issues as he did, before finally realizing that their own national agendas gave them different perspectives. Once he realized this, however, he realistically accepted the limitations thus evident: that the Russians and Turks might be allies, but they were not going to place their naval forces under anyone else's command—even his—and he was the star of the moment with a radiant reputation!

He had better fortune with the Sicilians and Portuguese, who did indeed entrust their squadrons to his command. After a shaky start and an appalling error of judgment regarding Caracciolo, he did a fine job as a coalition theater commander. Even when his allied subordinates let him down or performed poorly, he kept his anger or frustration limited to complaints in private letters to uninvolved persons. Outwardly he maintained an air of patience and collegiality and looked for ways to help his deficient friends improve without making them feel unwanted or inferior.

All things considered, Nelson was a model coalition commander: restrained, assuring, sympathetic, culturally aware, and sensitive to the needs, hopes, and fears of his partners. His relationship with Niza, which had all the potential for disaster and possibly even for a split between Britain and Portugal, is actually a paradigm of coalition collegiality and effectiveness. He worked hard with Niza—as partners and then as friends—to raise the performance of the Portuguese squadron without ever disrespecting or bullying Niza or his men. Nelson's relationship with Kadir Bey would, the evidence suggests, have proven as mutually rewarding had Ushakov's presence and personality not impeded its development.

Nelson certainly liked Kadir Bey and his nation. The Englishman's undisguised respect for the sultan of Turkey, his empire, and the Islamic faith did much to enhance Anglo-Turkish relations. It would prove equally rewarding in 1803, when Nelson returned to the Mediterranean and found himself able to use all the knowledge and experience he had gained back in 1798, 1799, and 1800. He would receive control of few allied vessels in the 1803 to 1805 period. Yet his regional awareness and his ability and willingness to see other national viewpoints, adopt conciliatory policies, and set common priorities helped him to cope well with a most difficult logistical and operational situation. We can only conclude that, despite his burning patriotism and excitable nature, Nelson performed very well in multinational situations.

Conclusion

LORD NELSON S colorful, complex character and supremely successful naval career are certainly the stuff of legend, but they are also of genuine importance for anyone trying to make sense of war and its nature as well as warriors and their natures. Nelson's positive qualities are so compelling and magnetic that not only do they seem to overwhelm his negative traits and render them less significant, but they also make the diminutive battle-maimed admiral a timeless model of martial excellence. Members of all armed services —soldiers, airmen, seamen, and marines—can learn much from a study of Nelson's way of war. He is a sterling example of how, even in organizations that place their greatest emphasis on uniform behavioral patterns and carefully choreographed collective action, individuals do matter. Those who extol the values of honor, valor, dedication, pride, and responsibility will perform to a high standard, shine among their peers, and ultimately earn the respect and gratitude of others.

I am clearly not alone in thinking so. In October 1998 United States secretary of the navy John H. Dalton explained in a Trafalgar Night speech at the Marine Barracks in Washington, D.C., how he had been grieved by the U.S. Navy's tarnished reputation in the early 1990s.[1] As part of a team dedicated to restoring the reputation and pride of U.S. seamen and marines, Dalton decided it was "time to take a good hard look at ourselves and the culture in which our people were serving and leading." One of his goals was to "reinvigorate our core values of Honor, Courage and Commitment, to once again give our sailors and marines the pride in themselves and pride in their actions that more properly reflect the legacy they inherited." It was a challenging task, Dalton explained, but he felt convinced it was both necessary and ultimately successful. "Today," he concluded, "our sailors and marines are respected as never before."

For those wondering what this has to do with Nelson, Dalton's explanation of the method he and his team developed and used is most illuminating:

> I began my own efforts in this endeavor by looking to the people and events that had inspired me, and had defined for me what our core values are. The examples I found were profoundly useful and satisfying, and it is those examples I will share with you tonight, because they all seem to come full circle to a familiar place . . . England, and can be traced to a British naval hero named Admiral Horatio Lord Nelson.[2]

This high praise from an American, and a secretary of the navy no less, is far more than just a tribute to a hero from a long-ago era; it is a tangible sign that Nelson's life and achievements do have relevance to today's armed forces.

Like his contemporary Wellington, Nelson committed himself wholeheartedly to his nation's service. His famous signal before Trafalgar encouraged his seamen to remember their "duty," and no word better sums up Nelson's entire life. By 1803 his duty included more than thirty years of active service for his nation. He had already masterfully won two major naval battles of his own and contributed significantly to several others. He had lost an arm, the sight of an eye, many teeth, and had sustained innumerable other wounds and injuries to his head and body in the many engagements and battles he fought. He was now gravely tired and wanted little more than to live out his remaining years in the peaceful company of Emma Hamilton and their daughter. He had achieved all the glory he had ever wanted—and he had always wanted plenty—and he knew that he was already the nation's paramount hero and could attain no greater fame, except perhaps through death in battle.

The Swiss writer Henri Frédéric Amiel once wrote that "heroism is the brilliant triumph of the soul over flesh." Few men in history have epitomized this more than Nelson. When his nation called him back to war in May 1803, the exhausted and badly maimed admiral never hesitated. His overwhelming sense of duty kept him continuously at sea for a further two years, during which time he reportedly never set foot on land, and after which he fought the greatest battle of his career—Trafalgar. Although certainly not suicidal, he adopted tactics that he knew contained dreadful personal risks. Yet, acutely aware of his responsibility to endure the same circumstances as his crews, he never hid himself away from those risks. His death at the moment of his greatest victory reveals the undeniable truth in both Amiel's statement and one of Nelson's last deathbed utterances: "Thank God, I have done my duty."

Nelson's patriotism, the source of his sense of duty, was always of the healthiest sort. It rested upon a firm belief that his society and its guiding ideas were established in accordance with God's principles, and that they best met the needs of most citizens. His patriotism also rested upon genuine love for his family, his community, and his people, and a sincere belief that they needed to be protected. It was never based on a jingoistic belief in Britain's cultural, colonial, or military superiority, or on a notion that the British had any right to dominate or exploit others. Indeed, his attitude toward his beloved homeland and its citizens' obligation to serve and defend it is in some ways akin to that encouraged in the early 1960s by the young President Kennedy, who urged the American public: "Ask not what your country can do for you, but what you can do for your country." This type of patriotism underpins, or at least should underpin, the motives of those who serve in defense forces.

Nelson matched his commitment to his nation with equal dedication to his navy; and that primarily meant his crews. He respected and obeyed his superiors (usually), but he *loved* his subordinates. He cared about them deeply and was unusually modern (in today's terms) in the attention he paid to their physical, emotional, and spiritual needs. He instinctively understood the highest principle of leadership: accomplish the mission *and* take care of your people.

As obsessive as Nelson was in his patriotic zeal and his pursuit of glory through victory, he never forgot that he relied on a vast number of men, who at best received a fraction of his own rewards and whose conditions of service were not conducive to their happiness and self-actualization. He therefore did his best to ensure that they felt wanted, trusted, and valuable. His efforts paid off. Wellington once said that Napoleon's presence on the battlefield was the equivalent of forty thousand men. He might also have been speaking of Nelson, whose effect was equally magical (and not just in terms of his well-known tactical brilliance and personal courage, of which he undoubtedly possessed plenty). His very presence in a fleet charged it with a psychological energy that has often been described but seldom adequately explained. Realizing that they served a commander who cared intensely about their well-being and would not ask them to do anything he had not already done himself, crews repaid Nelson's trust and care with great effort, pride, and valor. Their *élan* and *esprit de corps* soared. One sailor wrote home something that countless felt and discussed among themselves: "Men adored him. And in fighting under him, every man thought himself sure of success."

Having said all this, one must avoid falling into the trap of uncritical hero worship, of seeing only the admiral's wondrous feats and accomplishments and turning a proverbial Nelsonian blind eye to his failings. Another distinguished American leader, United States chief of naval operations Adm. Jay L. Johnson, recently stated during his own Trafalgar Night speech:

> We gather tonight not to celebrate victory, although to do so would certainly be appropriate, but to remember a man: Admiral Lord Nelson. When I consider this, though, it seems odd to differentiate between Nelson and victory. Are they not synonymous? Was it not also entirely appropriate for his flag to fly for the final time in a ship so named? The two are indistinguishable.[3]

Many books have been written that express nothing but the same sentiments. Yet comments of this uncritical eulogistic nature actually misrepresent Nelson, his accomplishments, and in particular, his relevance to today's military theorists and practitioners. Nelson was not always effective, let alone victorious. On some occasions he proved merely adequate, such as during the expedition up the San Juan River in 1780 and during the sieges of Bastia and Calvi in 1794. Try as he might, he never mastered land warfare, and he made a rather mediocre soldier. Because of his own biases, preconceptions, and service prerogatives, he made an even worse joint land-sea warrior. On several other occasions, even in the naval realm in which he ordinarily excelled, he failed miserably. His 1797 attack on Santa Cruz and his July 1801 raids on Boulogne were amateurishly or rashly conceived, inadequately resourced, and poorly directed. They hardly fit with Admiral Johnson's view that Nelson's life was an unbroken series of victories.

But Nelson's failings are, in fact, an integral part of his true significance to today's students of warriors and warfare. They reveal that humans can suffer tremendous personal and professional setbacks—including defeats and serious physical injury—or prove mediocre or even inadequate in some spheres of activity, but still make unique and outstanding contributions that are personally satisfying and of great significance to others. Nelson was no superman. Yet he persevered, tried to learn from his mistakes, experimented with techniques, tactics, and operational art, and remained true to himself and his ethical worldview.

His failings were not only professional; they were also personal. Often puffed up with pride, he was prone to dogmatism. Once he formed an opinion, almost nothing or no one (except perhaps his own hero, St. Vincent) could get him to alter it. This character trait led him to make some very poor choices, such as being excessively legalistic and consequently alienating locals in the Leeward Islands during the mid-1780s, having the unfortunate Commodore Caracciolo strung up from the yardarm of the frigate *Minerva* in July 1799, and refusing Admiral Keith's orders to send as many ships as possible to Minorca later that same month.

The flip side of this extreme and often harmful self-confidence, however, comprises qualities that deserve emulation by every serving member of the armed forces: *initiative* and *moral courage*. Nelson had these in abundance, and all three of his own great battles, along with others to which he decisively contributed, resulted from his willingness to act swiftly according to his own sense of opportunity and to do so regardless of traditions or instructions.

In today's world, and particularly in today's armed forces, initiative is certainly extolled. It forms a key component of the maneuverist warfighting doctrines currently in vogue throughout the Western world. Moral courage—the willingness to stand up for what you know to be right despite hierarchical or peer conformist pressures—is also a most desirable trait.

Yet individualism, self-assurance, and vanity, the very traits that make initiative and moral courage possible and instinctive, are seldom tolerated and often crushed in today's defense forces. More so now than even, for example, during World War II, when a few talented but unconventional officers still held command, defense forces have imposed a behavioral and character uniformity upon commanders that leaves little room for unusual or larger-than-life personalities. Almost certainly someone like Nelson, with supreme self-confidence, upright (and sometimes uptight) ethics, insubordinate habits, and a scandalous lifestyle (not to mention severe physical disabilities and lifelong hypochondria), would never rise in the services to take senior command today.

Our armed forces might do well to tolerate and even cultivate a little diversity and difference, throughout all levels, and to reduce their emphasis on regularity and conformity. Creative, profoundly religious, or eccentric personalities—even highly strung, passionate ones like Nelson's, Rommel's, and Patton's—have given frequent headaches to their superiors but also stunning victories and uplifting joy to their nations. Perhaps recognizing this, Caspar Weinberger, former United States secretary of defense, once said that "moral leadership, like Nelson's, must begin with confidence in ourselves. . . . We still revere Lord Nelson, not in spite of his boldness and insubordination, but *because* of it."[4] He added that Nelson's personality, "ardent, unruly, supremely self-confident," not only gave him unsurpassed success but also helped forge his nation's character.

Even if we disregard victories for the time being, as they imply battles and large-scale conflict, and think instead of peace support missions—a common feature of conflict during the last twenty or so years—one can still gain value from studying Nelson's career. Along with a warrior's heart, he possessed, paradoxically, a clergyman's conscience. He remained acutely concerned at the physical and psychological impact of war on both his own crews and all civilians. He was ever mindful of the need for adequate food, warmth, and other requirements for their happy existence. Also, despite a popular misconception that he was xenophobic—a claim trotted out but poorly argued in several recent biographies—he was actually interested in and sensitive to other people's cultural traditions, beliefs, and customs. Although a devout Protestant, he got on remarkably well with the Catholics he worked with, such as Niza, or those he served, such as Ferdinand and Maria Carolina. And he was never anything but thoughtful, helpful, and culturally sensitive in his dealings with Islamic leaders.

Although this book has much to say about warfighting, it is not intended only for real or armchair warriors. If I had any particular moral or practical lessons in mind when I wrote it, and I don't think I did, it may have been something to do with the strength of the human spirit. Its superiority over the frailty of our physical powers is quite wondrous. The story of Vice Admiral

Lord Nelson is uplifting. He endured frightful physical injuries, several defeats and stalemates, and much emotional trauma (some of it doubtless his own creation). Yet, through his strength of personality and highly unique warfighting style, he still managed to win three of naval history's greatest battles and earn a hero's immortality.

It is the triumph of Nelson's spirit as much as his tactical brilliance and physical and moral courage that places him among history's greatest warriors —that small group of exceptional humans who trusted their instincts, managed their fears, took risks, learned from mistakes, acted audaciously, and consequently proved all their foes unequal.

Notes

Abbreviations Used in the Notes

DLLN Nicolas, N. H., ed. *The Dispatches and Letters of Vice Admiral Lord Viscount Nelson, with Notes by Sir Nicholas Harris Nicolas, GCMG.* 7 vols. 1844–1846. Reprint, London: Chatham, 1998.

DSJM Maurice, J. F., ed. *The Diary of Sir John Moore.* 2 vols. London: Edward Arnold, 1904.

LESV Bonner-Smith, D., ed. *Letters of Admiral of the Fleet the Earl of St. Vincent whilst First Lord of the Admiralty, 1801–1804.* Vols. 55, 61. London: Navy Records Society, 1922, 1926.

MHNP Morrison, A., ed. *The Collection of Autograph Letters and Historical Documents Formed by Alfred Morrison (Second Series, 1882–1893), The Hamilton and Nelson Papers.* Vol. 1, *1756–1797.* Vol. 2, *1798–1815.* Privately printed 1893, 1894.

NCL Dawson, W., ed. *The Nelson Collection at Lloyds: A Description of the Nelson Relics and a Transcript of the Autograph Letters and Documents of Nelson and His Circle and of Other Naval Papers of Nelson's Period.* London: Macmillan for Lloyds, 1932.

ND *The Nelson Dispatch: Journal of the Nelson Society.*

NLW Naish, G., ed. *Nelson's Letters to His Wife and Other Documents 1785–1831.* London: Routledge and Kegan Paul in conjunction with the Navy Records Society, 1958.

NNJ Gutteridge, H., ed. *Nelson and the Neapolitan Jacobins: Documents Relating to the Suppression of the Jacobin Revolution at Naples, June 1799.* Vol. 25. Navy Records Society, 1903.

PCALC Hughes, E., ed. *The Private Correspondence of Admiral Lord Collingwood.* London: Navy Records Society, 1957.

Short titles are used for most sources in the notes. Full references may be found in the bibliography. Letters from Nelson generally do not include his name.

Chapter 1. Nelson's Conception of His Enemies

1. To His Highness the Bashaw of Tripoli, 28 April 1799, *DLLN,* 3:338.
2. Hibbert, *Nelson,* 182, 307.

3. To His Excellency Hugh Elliot, 8 October 1803, *DLLN,* 5:238.

4. Maccunn, *Contemporary English View,* 57–59.

5. Collingwood to Dr. Alexander Carlyle, 16 December 1798, *PCALC,* 92.

6. The best account of this period is found in Pocock, *Young Nelson.*

7. Collingwood to his sister, 22 March 1776, *PCALC,* 2.

8. To William Locker, 21 October 1781, *DLLN,* 1:48.

9. In later letters to friends, Nelson remarked bemusedly that he had once unknowingly detained the comte de Deux Ponts, commander of French forces at Yorktown, yet Nelson added no further comment on the battle or the loss of the colonies. Cf. *DLLN,* 1:90.

10. A valuable study of this period of Nelson's career is Kirby, "Nelson and American Merchantmen," 137–47. See also Oman, *Nelson,* 60, and following pages.

11. Collingwood to his sister, 2 January 1785, *PCALC,* 16–18; Kirby, "Nelson and American Merchantmen," 138.

12. Hibbert, *Nelson,* 50.

13. Letters to William Locker, 23 November 1784 and 15 January 1785, *DLLN,* 1: 112–14.

14. To William Locker, 15 January 1785, *DLLN,* 1:114.

15. To William Locker, 4 September 1785, *DLLN,* 1:138.

16. To Mrs. Frances ("Fanny") Nisbet, 19 August 1786, *NLW,* 34.

17. To Fanny, 17 April 1786, *NLW,* 28; *DLLN,* 1:166.

18. To Fanny, 23 April 1786, *NLW,* 30.

19. To Lord Sydney, 4 February 1786, *DLLN,* 1:153.

20. To Fanny, 4 May 1786, *NLW,* 32.

21. Ireland, *Naval Warfare,* 80–81, 85–87.

22. To William Locker, 12 August 1779, *DLLN,* 1:32.

23. See Nelson's own comment on the value of having French fluency during peacetime: Rawson, *Nelson's Letters,* 65. As for Nelson's continued efforts to master French himself, see Coleman, *Nelson,* 53, 74, 368 n. 10.

24. To William Locker, 2 November 1783, *DLLN,* 1:84.

25. To the Reverend Mr. Nelson, 10 November 1783, *DLLN,* 1:88.

26. To the Reverend Mr. Nelson, 31 January 1784, *DLLN,* 1:99.

27. To the Reverend Mr. Nelson, 28 December 1783, *DLLN,* 1:93.

28. To William Locker, 14 May 1784, *DLLN,* 1:107.

29. Warner, *Portrait,* 34.

30. To Fanny, 7 January 1793, *NLW,* 72.

31. Dickinson, *Britain and the French Revolution,* 128–29.

32. Ibid., 130.

33. Collingwood to Dr. Alexander Carlyle, 27 September 1792, *PCALC,* 33.

34. To William Locker, 10 September 1789, in Rawson, *Nelson's Letters,* 64.

35. Rose, *Lord Hood.*

36. Connelly, *Blundering to Glory*, 16–18.
37. McLynn, *Napoleon*, 75; Schom, *Napoleon Bonaparte*, 22.
38. To Fanny, 7 September 1793, *NLW*, 89.
39. McLynn, *Napoleon*, 76; Schom, *Napoleon Bonaparte*, 22.
40. Sir Sidney Smith repeated an even more fanciful figure of 120,000. Smith to Sir William Hamilton, 24 December 1793, in *MHNP*, 1:183–84.
41. To Fanny, 23 December 1793, *NLW*, 97.
42. To the Reverend Edmund Nelson, 20 August 1793, *DLLN*, 1:320.
43. To Fanny, 4 August 1793, *NLW*, 86.
44. Cf., "During the whole day the miserable and terrified inhabitants were crowding into boats with children of all ages to escape the death that waited them on shore. . . . The scene has been shocking." From Elliot, *Life and Letters*, 2:200–201; On 18 February 1794, for instance, Lieutenant Bowen told Sir William Hamilton that he couldn't wait to repay "the rascals for their True Conduct at Toulon" (*NCL*, 50).
45. To HRH the Duke of Clarence, 27 December 1793, *DLLN*, 1:343, 344.
46. To Fanny, 27 December 1793, *NLW*, 97; *DLLN*, 1:345.
47. Collingwood to Sir Edward Blackett, 2 March 1794, *PCALC*, 43.
48. To Fanny, 27 September 1793, *DLLN*, 1:487.
49. To William Locker, 10 October 1794, *DLLN*, 1:491.
50. Collingwood and others felt the same frustrations, and in moments of depression also thought the French should be left to slaughter themselves. See Collingwood to Dr. Alexander Carlyle, 9 July 1794, *PCALC*, 51–53.
51. To Sir William Hamilton, 7 May 1794, *MHNP*, 1:191.
52. To Fanny, 25 February 1795, *DLLN*, 2:8.
53. To Fanny, 1 April 1795, *NLW*, 204.
54. Ibid.; to Fanny, 25 February 1795, *DLLN*, 2:26.
55. To Fanny, 1 July 1795, *DLLN*, 2:49.
56. To Sir William Hamilton, 7 April 1795, *MHNP*, 1:206.
57. To William Suckling, 27 October 1795, *DLLN*, 2:92–93.
58. Ibid.
59. To the Reverend Dixon Hoste, 12 December 1795, *MHNP*, 1:217.
60. Ibid.; italics added for emphasis.
61. Connelly, *Blundering to Glory*, 23–49.
62. To William Suckling, 27 July 1795, *DLLN*, 2:62.
63. Nelson memoranda, *DLLN*, 2:198. Cf. Collingwood to Sir Edward Blackett, 27 August 1796, *PCALC*, 76–78.
64. Letter to the Right Honorable Gilbert Elliot, 5 July 1796, *DLLN*, 2:203.
65. To Capt. Cuthbert Collingwood, 1 August 1796, *MHNP*, 1:222.
66. Letter to Adm. Sir John Jervis, 18 July 1796, *DLLN*, 2:217.
67. White, *1797*, 15.
68. To Sir William Hamilton, 1 December 1796, *MHNP*, 1:226.

69. To Fanny, 13 October 1796, *NLW,* 305.

70. White's marvelous account of this ill-fated mission is the best to date. See note 67 above.

71. An excellent firsthand account of the attack on Tenerife can be found in *Wynne Diaries,* Anne Fremantle, ed.

72. Minutes of Lord Grenville, 28 April 1798, *NCL,* 292; Minutes of Orders to Lord St. Vincent, 28 April 1798, *NCL,* 292.

73. Tracy, *Nelson's Battles,* 106. The most thorough analysis of St. Vincent's decision is found in Orde, *Mediterranean Command.*

74. Lavery, *Nelson and the Nile,* 131, 133.

75. Cf. To His Royal Highness the Duke of Clarence, 24 April 1798, *DLLN,* 3:10.

76. To the Earl of St. Vincent, 17 May 1798, *DLLN,* 3:16.

77. To Fanny, 20 July 1798, *NLW,* 398.

78. To the Right Honorable Sir William Hamilton, 17 June 1798, *MHNP,* 2:11.

79. To George Baldwin, 24 June 1798, *DLLN,* 3:37.

80. To the Right Honorable Sir William Hamilton, 20 July 1798, *DLLN,* 3:42.

81. Lavery, *Nelson and the Nile,* 120.

82. To the Right Honorable Sir William Hamilton, 14 June 1798, *DLLN,* 3:30.

83. To Fanny, 11 August 1798, *NLW,* 399.

84. To Fanny, 28 September 1798, *NLW,* 402.

85. To the Honorable William Wyndham, 21 August 1798, *DLLN,* 3:109.

86. Cf. the entries of 2 and 3 October 1798 in Fremantle, *Wynne Diaries;* Macleod, *War of Ideas,* 4, 49–50, 185.

87. Hibbert, *Nelson,* 163–65.

88. Tracy, *Nelson's Battles,* 123; Macleod, *War of Ideas,* 4, 49–50.

89. At one point he even claimed to be as committed to Ferdinand's cause as he was to King George's. Letter to Sir John Acton, 7 March 1799, in *NNJ,* 7.

90. To Vice Admiral Goodall, 31 January 1799, *DLLN,* 3:246; To Captain Ball, 31 January 1799, *DLLN,* 3:247.

91. To Sir John Acton, 12 March 1799, *NNJ,* 8.

92. The best collection of primary documents on this strange period in Nelson's life undoubtedly remains *NNJ.*

93. To Mrs. Cadogan, 17 July 1799, *MHNP,* 2:55; Lavery, *Nelson and the Nile,* 280.

94. Parsons, *Nelsonian Reminiscences,* 3.

95. Ibid., 5.

96. See Nelson's opinion, 26 June 1799, *NNJ,* 217.

97. Captain Ball to Lady Hamilton, 5 February 1799, *MHNP,* 2:37.

98. To Captain Troubridge, 18 February 1799, *DLLN,* 3:268.

99. To Sir John Acton, 2 May 1799, *NNJ,* 14.

100. To Rear Admiral Duckworth, 25 June 1799, *NNJ,* 216.

101. Nelson's advice to the king, 11 July 1799, *NNJ,* 306. Italics added for emphasis.

102. To Lord Spencer, 13 July 1799, *NNJ,* 308; letters to Vice Admiral Lord Keith, 14 and 19 July 1799, *DLLN,* 3:411, 414.

103. *DLLN,* 3:410 n.; Minutes of Instructions to Lord Nelson, 20 August 1799, *NCL,* 322–23.

104. To Abdul Cadir Bey, 9 March 1799, *DLLN,* 3:288.

105. To HRH the Duke of Clarence, 11 April 1799, *DLLN,* 3:324.

106. To Sir Thomas Troubridge, 19 August 1799, *DLLN,* 3:451.

107. To Admiral the Earl of St. Vincent, 20 January 1801, *DLLN,* 4:275.

108. To Evan Nepean, 15 June 1801, *NCL,* 171.

109. To Emma Hamilton, 16 March 1801, *ND* 7, pt. 6 (April 2001): 385.

110. To Sir Hyde Parker, 24 March 1801, in Rawson, *Nelson's Letters,* 320.

111. *DLLN,* 4:310.

112. Capt. Thomas Fremantle to the Marquis of Buckingham, 6 April 1801, in Fremantle, *Wynne Diaries,* 3:47–48.

113. To Emma, 9 April 1801, in Rawson, *Nelson's Letters,* 326.

114. And Nelson certainly did inspire this great public confidence. Cf. the Earl of St. Vincent letters to Nelson, 3 August 1801 and 8 August 1801, in *LESV,* 1:130, 133.

115. To the Earl of St. Vincent, 6 August 1801, *DLLN,* 4:445.

116. To the Earl of St. Vincent, 11 August 1801, in Rawson, *Nelson's Letters,* 348.

117. To the Squadron, 18 August 1801, in Rawson, *Nelson's Letters,* 350.

118. William Cathcart to Lord Cathcart, 20 August 1801, in Laughton, *Naval Miscellany,* 1:295.

119. Macleod, *War and Ideas,* 194–95.

120. To the Earl of St. Vincent, October [3?] 1801, *DLLN,* 4:505.

121. Admiral the Earl of St. Vincent to William Baker, 5 October 1801, *LESV,* 1:284; Admiral the Earl of St. Vincent to the Earl of Uxbridge, 12 October 1801, *LESV,* 1:285; Elliot, *Life and Letters,* 3:259.

122. Cf. Durant, "Nelson's Visit to Monmouth," vol. 4.

123. Macleod, *War and Ideas,* 194.

124. To Dr. Baird, 11 October 1801, *DLLN,* 4:508.

125. To Dr. Baird, 24 September 1801, *NCL,* 278.

126. As did his friend Collingwood, who wrote that "a Frenchman is such a trickish animal that I cannot consider him but with suspicion, and a Corsican has given proof that he can out-trick a Frenchman." Collingwood to Sir Edward Blackett, 28 January 1802, *PCALC,* 137.

127. Cf. Capt. William Johnstone Hope to Capt. Benjamin Hallowell, 19 November 1801, *NCL,* 174.

128. Even before the October 1801 armistice, he had privately insisted that no peace with Napoleon's government could last, but would merely be an interregnum. Cf. To Sir Alexander Ball, 4 June 1801, *NCL,* 166.

129. To Sir Brooke Boothby, 1 May 1802, *DLLN,* 5:12.

130. Herold, *Mind of Napoleon,* 125.

131. Macleod, *War and Ideas,* 4, 204.

132. Maccunn, *Contemporary English View,* 66.

133. Collingwood to Dr. Alexander Carlyle, 27 August 1804, *PCALC,* 156.

134. The Earl of St. Vincent to Lord Nelson, 19 May 1803, *LESV,* 2:316–17.

135. Morriss, *Nelson Life and Letters,* 134.

136. To Admiral the Earl of St. Vincent, 27 September 1803, *DLLN,* 5:214.

137. Others, including Collingwood, occasionally expressed grudging praise for Napoleon's cunning and genius for war. Cf. Collingwood to Sir Edward Blackett, 27 August 1796, *PCALC,* 78.

138. To the Grand Vizir, 11 February 1804, *DLLN,* 5:414.

139. Cf. my article, "Nelson's Warfighting Style," 16–37. Collingwood, doubtless inspired by his friend's expressed desires for "annihilation" battles, himself wrote on one occasion that "extermination" of the French should be attempted. Collingwood to Dr. Alexander Carlyle, 17 June 1804, *PCALC,* 154.

140. Wybourn, *Sea Soldier,* 77.

141. To Sir Alexander John Ball, 19 April 1805, *DLLN,* 6:410.

142. To Sir Alexander Davison, 6 April 1805, *DLLN,* 6:400.

143. To Viscount Sidmouth, 11 May 1805, *DLLN,* 6:436.

144. Cf. Sarah Connor to Lady Hamilton, 20 August 1805, *NCL,* 212.

145. To the Grand Vizir, 13 June 1804, *DLLN,* 6:71.

Chapter 2. The Admiral's Spiritual Beliefs

1. *DLLN,* 7:139–40; Warner, *Nelson's Last Diary,* 28, 70, 71.

2. "Sketch of My Life," 15 October 1799, *DLLN,* 1:3–15.

3. A rich source of information on the Nelsons and their familial interactions is Matcham's *Nelsons of Burnham Thorpe.*

4. Scott, *Recollections,* 191.

5. Warner, *Nelson's Last Diary,* 10–11. Nelson's statement could equally make sense without the insertion of "I," as he trusted his care all day to the Lord.

6. Clarke and M'Arthur, *Life of Nelson,* 20.

7. To Lieutenant Governor Locker, 9 February 1799, *DLLN,* 3:260.

8. Ibid.

9. To John Locker, 27 December 1800, *DLLN,* 4:271.

10. To George Smith, 31 August 1801, *DLLN,* 4:480–81.

11. To Captain Locker, 12 August 1777, *DLLN,* 1:23.

12. Cf., Captain Hardy's comment: "Providence has been so good to me since my departure from Spithead that I have no disasters to relate." Nelson to Mr. Manfield, 29 May 1802, in Broadley and Bartelot, *Nelson's Hardy,* 92; Wybourn, *Sea Soldier,* 71.

13. To Lieutenant Governor Locker, 21 March 1795, *DLLN,* 2:21. Italics added for emphasis.

14. To Fanny, 10 March 1795, *NLW,* 199.

15. To Fanny, 11 September 1785, *NLW,* 19.

16. To Fanny, 3 March 1786, *NLW,* 22–23; *DLLN,* 1:139.

17. Nelson's naval peers, although not as religious as their famous colleague, had similar political attitudes. After visiting a house adorned with pictures of Paine and others, Collingwood complained that the owners were "strange characters." Adm. Cuthbert Collingwood to Sir Edward Blackett, 4 May 1799, *PCALC,* 98.

18. To Commissioner Otway, 24 November 1804, *DLLN,* 6:280.

19. Vice Admiral Vernon to Mrs. Vernon, 31 March 1741, in Laughton, *Naval Miscellany,* 1:393.

20. To Lindholm, 12 April 1801, and Nelson to Sneedorf, 12 April 1801, *ND* 7, pt. 6 (April 2001): 394. My thanks to Colin White.

21. Jennings, *Croker Papers,* 2:233–34.

22. Gash, *Wellington Anecdotes,* 16.

23. To Fanny, 1 April 1795, *NLW,* 204.

24. To Fanny, 8 July 1794, *NLW,* 117.

25. *DLLN,* 2:340–43.

26. Coleman, *Nelson,* 128–29.

27. To Fanny, 20 July 1798, *NLW,* 398.

28. To Emma, 18 February 1800, *MHNP,* 2:86. For Nelson's quotations from Shakespeare, see White, "Nelson and Shakespeare," 145–50.

29. Cf. Elliot, *Life and Letters,* 3:363.

30. Ibid., 147.

31. *DSJM,* 1:367.

32. Lord St. Vincent to Evan Nepean, 17 January 1801, in Laughton, *Naval Miscellany,* 2:332; Hibbert, *Nelson,* 237.

33. To Fanny, 27 June 1794, *NLW,* 115.

34. To Fanny, 28 June 1794, *NLW,* 116.

35. To Sir Evan Nepean, 26 December 1803, *NCL,* 183.

36. The Earl of St. Vincent to Nelson, 14 July 1797, *DLLN,* 2:413.

37. The two best works on the Tenerife campaign are Guimerá, *Nelson and Tenerife* and White, *1797,* cited in chapter 1, note 67. An excellent firsthand account of the attack can be found in Fremantle, *Wynne Diaries.*

38. To Capt. Sir Andrew Snape Hamond, 8 September 1797, *DLLN,* 2:443. Italics added for emphasis.

39. Cf. To Admiral the Earl of St. Vincent, 18 September 1797, *DLLN,* 2:446.

40. To Fanny, 5 August 1797, *NLW,* 332.

41. To Fanny, 24 May 1798, *NLW,* 396.

42. To Admiral the Earl of St. Vincent, 31 May 1798, *DLLN,* 3:23.

43. Elliot, *Life and Letters,* 3:242.

44. To Fanny, 19 August 1786, *NLW,* 33.

45. Moorhouse, *Nelson in England,* 114.

46. Ibid., 115.

47. Knight, *Autobiography,* 1:139.

48. To Fanny, 4 March 1801, *MHNP,* 2:125.

49. Moorhouse, *Nelson in England,* 186.

50. To Emma, 8 February 1801, *DLLN,* 4:284.

51. To Emma, 11 March 1801, *MHNP,* 2:128.

52. To Emma, 3 February 1801, *MHNP,* 2:110.

53. To Emma, February 1801, *MHNP,* 2:113.

54. Even Hardy worried about Nelson's ability to make his relationship with Emma seem "honorable" following the death of Sir William. Letter to Mr. Manfield, 6 April 1803, in Broadley and Bartelot, *Nelson's Hardy,* 105.

55. Cf. To Emma, September 1801, *MHNP,* 2:165.

56. To Emma, 1 March 1801, *MHNP,* 2:123. Italics added for emphasis.

57. Russell, *Nelson and the Hamiltons,* 451.

58. To Emma, 9 March 1801, *MHNP,* 2:127.

59. Pocock, *Horatio Nelson,* 226.

60. To Emma, 10 March 1801, *MHNP,* 2:127–28.

61. To the Right Honorable Sir Gilbert Elliot, 5 August 1796, *DLLN,* 2:234; Nelson to Fanny, 29 June 1797, *NLW,* 327.

62. Moorhouse, *Nelson in England,* 90; Clarke and M'Arthur, *Life of Nelson,* 201.

63. To Fanny, 30 June 1797, *NLW,* 328.

64. Jordan, "Nelson as Popular Hero," 117.

65. To Emma, 6 October 1803, *MHNP,* 2:219.

66. Clarke and M'Arthur, *Life of Nelson,* 12.

67. See the statement by John Pasco, the flag lieutenant on the *Victory, DLLN,* 7:150.

68. To Fanny, 22 April 1794, *DLLN,* 1:385.

69. To the Right Honorable Sir William Hamilton, 18 June 1798, *DLLN,* 3:34.

70. To the Right Honorable Sir William Hamilton, 14 June 1798, *DLLN,* 3:30.

71. Moorhouse, *Nelson in England,* 87–88.

72. *DLLN,* 4:309.

73. Moorhouse, *Nelson in England,* 109.

74. Ibid., 160.

75. Warner, *Nelson's Last Diary,* 11.

76. Hibbert, *Nelson,* 269.

77. Private diary, 13 September 1805, *DLLN,* 7:33–35.

78. *DLLN,* 7:140; Hibbert, *Nelson,* 366.

79. Cf. To the Reverend Mr. Nelson, 8 February 1782, *DLLN,* 1:58.

80. Scott, *Recollections,* 190.

81. Ibid., 187.

82. To the Reverend Mr. Nelson, 20 February 1785, *DLLN,* 1:124–25.

83. Clarke and M'Arthur, *Life of Nelson,* 59.

84. *DLLN,* 1:124.

85. Finlayson, "Midshipman at Copenhagen," 88–89.

86. Ibid., 96, and Colonel Stewart's narrative, *DLLN,* 4:386.

87. *DLLN,* 1:124; Clarke and M'Arthur, *Life of Nelson,* 71.

88. Pocock, *Remember Nelson,* 38.

89. Ibid., 45.

90. Ibid., 62.

91. Scott, *Recollections,* 191.

92. Ibid., 192.

93. "Extract of a Letter from a gentleman of High Rank, dated on board his Majesty's ship, *Elephant* (the ship on board of which Lord Nelson was), Copenhagen Roads, April 3, 1801," *The Edinburgh Advertiser,* 24 April 1801, 261.

94. Memorandum to the Respective Captains of the Squadron, 2 August 1798, *DLLN,* 3:61.

95. Parsons, *Nelsonian Reminiscences,* 23.

96. Jordan, "Nelson as Popular Hero," 112.

97. Rodger, *Wooden World,* 45.

98. Shannon, "Missing Nelson Letters," 595–98.

99. For Nelson's letter to the SPCK, 4 March 1793, see *ND* 7, pt. 7 (July 2001): 446.

100. Lord Nelson to the Society for Promoting Christian Knowledge, 4 January 1801, in Allen and McClure, *Two Hundred Years,* 456.

101. To HRH the Duke of Clarence, 3 November 1792, *DLLN,* 1:293.

102. Hibbert, *Nelson,* 43.

103. To General Lindholm, 16 June 1801, *MHNP,* 2:156.

104. To Count Thurin, Commodore and Commander of His Sicilian Majesty's Frigate *La Minerva,* 29 June 1799, *DLLN,* 3:398; Nelson to His Majesty the King of the Two Sicilies, 13 August 1799, *DLLN,* 3:439.

105. Cf. To Cardinal Despuig, August 1804, *DLLN,* 6:182–83.

106. To Captain Hardy, 12 January 1799, *DLLN,* 3:231–32.

107. Don Antonio Biancareddu to Nelson, 22 October 1804, 36.

108. Letters to the Right Honorable Sir Gilbert Elliot and to Adm. Sir John Jervis, 18 July 1796, *DLLN,* 2:215–16; Nelson to the Honorable William Frederick Wyndham, 2 March 1799, *DLLN,* 3:277.

109. To His Holiness the Pope, 24 June 1800, *DLLN,* 4:259.

110. But see his comment in his "Sketch of My Life," 15 October 1799, *DLLN,* 1:15.

111. To Emma, 1 October 1805, *DLLN,* 7:60.

Chapter 3. Command, Leadership, and Management

1. *The Great Commanders,* "Nelson," produced in Great Britain by Channel 4 television.

2. The literature is vast, but a good starting point for analysis of military command is Sheffield, *Leadership and Command,* especially 1–14.

3. Lavery, *Nelson's Navy,* 216–17; Pope, *Life in Nelson's Navy,* 32, 64, 244–45, 250.

4. Clarke and M'Arthur, *Life of Nelson,* 23.

5. To Fanny, 27 January 1796, *NLW,* 282.

6. To Lord Sydney, Secretary of State, 20 March 1785, *DLLN,* 1:130–31.

7. To Rear Adm. Sir Richard Hughes, 15 February 1785, *DLLN,* 1:120.

8. To Fanny, 3 March 1786, *NLW,* 23.

9. Cf. Minutes of Instructions to Lord St. Vincent, October 1798, *NCL,* 306–7.

10. To Lord Spencer, 13 July 1799, *NNJ,* 308; Nelson letters to Vice Admiral Lord Keith, 13, 14, and 19 July 1799, *DLLN,* 3:408, 411, 414.

11. See *DLLN,* 3:352–66.

12. To Fanny, 24 July 1795, *NLW,* 217.

13. *DLLN,* 4:312.

14. Hibbert, *Nelson,* 43.

15. To Adm. Sir John Jervis, 30 May 1796, *DLLN,* 2:175.

16. A fine study is provided in Dugan, *Great Mutiny.*

17. To Fanny, 15 June 1797, *NLW,* 326.

18. To Fanny, 11 July 1797, *NLW,* 330.

19. To Sir Robert Calder, Knight, First Captain to Admiral the Earl of St. Vincent, 9 July 1797, *DLLN,* 2:409.

20. To Philip Stephens, Admiralty, 11 July 1787, *DLLN,* 1:246.

21. To Philip Stephens, Admiralty, 17 August 1787, *DLLN,* 1:253.

22. To the Respective Captains and Commanders of His Majesty's Ships and Vessels on the Mediterranean Station, 7 November 1803, *DLLN,* 5:284–85.

23. Dann, *Nagle Journal,* 209.

24. Ibid.

25. Lord St. Vincent to Evan Nepean, 9 November 1800, in Laughton, *Naval Miscellany,* 2:329.

26. Cf. Pocock, *Horatio Nelson,* 81; Coleman, *Nelson,* 5, 85, 89.

27. To an unnamed admiral, January 1804, *DLLN,* 5:385–86.

28. Ibid.

29. Hibbert, *Wellington,* 96, 139.

30. To the Reverend Mr. Nelson Jr., 18 October 1781, *DLLN,* 1:46.

31. To the Right Honorable Earl Spencer, 8 April 1800, *DLLN,* 4:224.

32. White, *1797,* 125; Hibbert, *Nelson,* 123.

33. Pocock, *Horatio Nelson,* 165; Tracy, *Naval Chronicle,* 1:260.

34. To Captain Locker, 12 July 1783, *DLLN,* 1:76.

35. Ibid.

36. To the Captains of the Ships of the Squadron, 2 August 1798, *DLLN,* 3:61.

37. To Admiral the Earl of St. Vincent, 3 August 1798, *DLLN,* 3:56–57.

38. To Earl Spencer, 7 September 1798, *DLLN,* 3:115.

39. To Dr. Moseley, Chelsea Hospital, 11 March 1804, *DLLN,* 5:438; Tracy, *Naval Chronicle,* 13.

40. Jordan, "Nelson as Popular Hero," 113.

41. To Commodore Linzee, 6 February 1794, *ND* 7, pt. 1 (January 2000): 34.

42. To Dr. Moseley, Chelsea Hospital, 11 March 1804, DLLN, Vol. V, 438; Tracy, *Naval Chronicle,* 13.

43. Berckman, *Nelson's Dear Lord,* 56.

44. Sheffield, *Leadership and Command,* 117.

45. An excellent book on command types (although I don't necessary agree with all conclusions) is Dixon, *On the Psychology of Military Incompetence.*

46. Fremantle to his wife Betsey, 1 October 1805, in *Wynne Diaries,* entry for that date.

47. To Vice Admiral Lord Keith, 8 April 1800, *DLLN,* 4:226.

48. Ibid.

49. To Lieutenant Governor Locker, Royal Hospital, Greenwich, 9 February 1799, *DLLN,* 3:260.

50. Fanny to Nelson, 11 March 1797, *NLW,* 352.

51. "A Narrative of the Services of Captain Richard [*sic*] Willett Miller of His Majesty's Ship *Theseus* (written to his wife) in the bombardment of Cadiz, and in the glorious tho' unsuccessful attack on the town on Santa Cruz in the island of Teneriffe in the month of July, 1797," in Buckland, *Miller Papers,* 9.

52. To Admiral Sir John Jervis, 4 July 1797, *DLLN,* 2:405; Tracy, *Naval Chronicle,* 1:183.

53. "Letter from a Gentleman, present at the Festivities at Fonthill, to a Correspondent in Town," dated 28 December 1800, transcribed in *ND* 7, pt. 4 (October 2000): 227–31.

54. Berckman, *Nelson's Dear Lord,* 153.

55. Ibid., 240.

56. Fitchett, *Nelson and His Captains,* 12.

57. *DLLN,* 4:308.

58. Ibid., 309; Fitchett, *Nelson and His Captains,* 104.

59. Tracy, *Naval Chronicle,* 1:187.

60. Berry, "An Authentic Narrative of the Proceedings of His Majesty's Squadron under the Command of Rear Admiral Sir Horatio Nelson, From its Sailing from Gibraltar to the Conclusion of the Glorious Battle of the Nile. Drawn up from the Minutes of an Officer of Rank in the Squadron," Tracy, *Naval Chronicle,* 1:246.

61. Ibid.

62. Lord St. Vincent to Evan Nepean, 9 November 1800, in Laughton, *Naval Miscellany,* 2:329.

63. To Mr. Charles Connor, December 1803, *DLLN,* 5:310–11.

64. Lewis, *Social History,* 202–4.

65. To Admiral the Earl of St. Vincent, 12 December 1803, *DLLN,* 5:308.

66. To Adm. Sir John Jervis, 31 May 1796, *DLLN,* 2:176–77.

67. To Admiral the Earl of St. Vincent, 11 January 1804, *DLLN,* 5:364.

68. To Captain Troubridge, HM Ship *Culloden,* 30 March 1799, *DLLN,* 3:310.
69. Pocock, *Horatio Nelson,* 319.
70. To Fanny, 2 August and 22 November 1796, *NLW,* 298, 309.
71. *DLLN,* 2:344.
72. Fitchett, *Nelson and His Captains,* 71.
73. To Admiral the Earl of St. Vincent, 3 August 1798, *DLLN,* 3:57.
74. To Sir Edward Berry, 31 August 1805, *DLLN,* 7:23.
75. To Admiral Lord Hood, 10 August 1794, *DLLN,* 1:477.
76. Letters to Fanny, 1 and 12 September 1794, *NLW,* 121, 122, 123.
77. To Fanny, 10 October 1794, *NLW,* 126.
78. Warner, *Portrait of Nelson,* 251.
79. To Admiral the Earl of St. Vincent, 8 July 1803, *DLLN,* 5:122.
80. Mahan, *Life of Nelson,* 572.
81. Admiral the Earl of St. Vincent to Lord Nelson, 21 August 1803, *LESV,* 2:320.
82. To the Right Honorable Henry Addington, 24 August 1803, *DLLN,* 5:174; Nelson to Admiral the Earl of St. Vincent, 27 September 1803, *DLLN,* 5:214.
83. To Admiral the Earl of St. Vincent, 27 September 1803, *DLLN,* 5:214.
84. For a useful source, see Scott, *Recollections.*
85. To Captain Moubray, 8 March 1804, *DLLN,* 5:437.
86. Winton, *Illustrated History,* 85.
87. Wybourn, *Sea Soldier,* 72.
88. Ibid., 77.
89. Gwyther. "Nelson's Gift," *Trafalgar Chronicle,* 10:47–56.
90. To His Excellency Hugh Elliot, 1 November 1803, in Rawson, *Nelson's Letters,* 398.
91. To Alexander Davison, 12 December 1803, *DLLN,* 5:306.
92. To Admiral the Earl of St. Vincent, 12 December 1803, *DLLN,* 5:307.
93. To John Tyson, 12 December 1803, *DLLN,* 5:309.
94. Cf. To Nathaniel Taylor, Naval Officer, Malta, 8 June 1804, *DLLN,* 6:65.
95. To the Commissioners of the Transport Board, London, 7 August 1804, *DLLN,* 6:138–39.
96. Forester, *Lord Nelson,* 285.
97. Mahan, *Life of Nelson,* 572–73.
98. To William Marsden, Admiralty, 2 October 1805, *DLLN,* 7:62–63.
99. To Lord Barham, First Lord of the Admiralty, 5 October 1805, *DLLN,* 7:75.
100. Oman, *Nelson,* 566.

Chapter 4. Nelson's Warfighting Style and Maneuver Warfare

1. *MCDP 1: Warfighting,* 73.
2. Cf. Bolger. "Maneuver Warfare Reconsidered," in *Maneuver Warfare,* ed. Hooker, 19–41.

3. Hooker, *Maneuver Warfare* and Leonhard, *Art of Maneuver.* Both authors draw inspiration from the "granddaddy" of maneuverist works: Simpkin's seminal *Race to the Swift.* They also refer frequently to another pioneering but less sturdy work: Lind's *Maneuver Warfare Handbook.* These studies also pay inadequate attention to warfare at sea.

4. Hooker, *Maneuver Warfare,* 10; Leonhard, *Art of Maneuver,* 114.

5. *MCDP 1–3: Tactics,* 30.

6. Clausewitz. *Vom Kriege,* 671–72.

7. Hooker, *Maneuver Warfare,* 10.

8. Maffeo, *Most Secret and Confidential,* 91–103.

9. To Viscount Castlereagh, 5 October 1805, *DLLN,* 7:76.

10. To Earl Spencer, 9 August 1798, *DLLN,* 3:98.

11. Adm. Cuthbert Collingwood to Dr. Alexander Carlyle, 24 August 1801, *PCALC,* 130.

12. Berry's narrative in *DLLN,* 3:50.

13. Several writers believe the bold maneuver was the spontaneous decision of Captain Foley of the *Goliath,* who moved inside the French fleet without waiting for Nelson's order. See Kennedy, *Band of Brothers,* 128; Bradford, *Nelson,* 200; Foreman and Phillips, *Napoleon's Lost Fleet,* 123–24.

14. I am especially persuaded by Captain Hood's letter to Lord Bridport shortly after the battle, published in *DLLN,* 3:63–64. See also Bennett, *Nelson the Commander,* 131; Pocock, *Horatio Nelson,* 163.

15. Foreman and Phillips, *Napoleon's Lost Fleet,* 124.

16. To Fanny, 12 April 1795, *NLW,* 206.

17. To Fanny, 14 March 1795, *NLW,* 200.

18. *DLLN,* 2:14.

19. Hooker, *Maneuver Warfare,* 10–13; Leonhard, *Art of Maneuver,* 113–18.

20. Patton. *War as I Knew It,* 338.

21. The best analysis of Nelson's command and control style is undoubtedly found in Palmer, "The Soul's Right Hand," 679–706.

22. He also made a close study of British naval tactics used against the American colonies during the War of Independence. Clarke and M'Arthur, *Life of Nelson,* 93.

23. White, *1797,* 21; *DLLN,* 2:311.

24. A fine study of Jervis's signals during the Battle of Cape St. Vincent can be found in Palmer, "Sir John's Victory," 31–46.

25. King, *Every Man Will Do His Duty,* 81; *DLLN,* 2:347.

26. Clarke and M'Arthur, *Life of Nelson,* 101.

27. Tracy, *Naval Chronicle,* 1:97.

28. White, *1797,* 105.

29. *DLLN,* 2:417.

30. Guimerá, *Nelson and Tenerife,* 7. This work and White's are the finest accounts of Nelson's darkest moments.

31. Lady Minto to Lady Malmesbury, 7 September 1800, in Elliot, *Life and Letters,* 3:150.

32. Weller, *Wellington,* 116, 117.

33. Herold, *Mind of Napoleon,* 218.

34. To Emma, 1 October 1805, *DLLN,* 7:60.

35. Tracy, *Naval Chronicle,* 1:10.

36. *DLLN,* 7:89–92.

37. To Vice Admiral Collingwood, 9 October 1805, *DLLN,* 7:95.

38. To Rear Admiral Duckworth, 20 August 1799, *DLLN,* 3:453–54.

39. Ibid., 453.

40. Ibid., 454.

41. Wybourn, *Sea Soldier,* 75.

42. Ibid.

43. To the Right Honorable Earl Spencer, 6 November 1799, *DLLN,* 4:90.

44. Hooker, *Maneuver Warfare,* 63, 80, 81; Leonhard, *Art of Maneuver,* 18.

45. Thursfield, *Nelson and Naval Studies,* 16; Nelson to William Locker, 25 February 1783, *DLLN,* 1:72.

46. Lind, "Misconceptions of Maneuver Warfare," *Marine Corps Gazette* 72, no. 1 (January 1988): 17.

47. To Adm. Sir Hyde Parker, 24 March 1801, *DLLN,* 4:297–98.

48. *MCDP 1: Warfighting,* 73.

49. *PCALC,* 92.

50. Ibid., 130.

51. To Fanny, 1 April 1795, *NLW,* 204.

52. *MCDP 1–3: Tactics,* 57–78.

53. Van Creveld, *Air Power,* 3; Leonhard, *Art of Maneuver,* 51, 87–88.

54. Berry, "An Authentic Narrative . . . ," in Tracy, *Naval Chronicle,* 1:243.

55. *LESV,* 1:64. Italics added for emphasis.

56. To Sir Thomas Troubridge, 29 March 1801, in Laughton, *Naval Miscellany,* 1:424.

57. Letter to the Right Honorable Earl Howe, 8 January 1799, *DLLN,* 3:230.

58. *DLLN,* 7:241 n.

59. Tracy, *Naval Chronical,* 1:13; Terraine, *Trafalgar,* 51.

60. To Fanny, 1 April 1795, *NLW,* 204.

61. *NLW,* 204; *DLLN,* 2:26.

62. To Alexander Davison, 6 September 1805, *DLLN,* 7:30.

63. To Alexander Davison, 30 September 1805, *DLLN,* 7:56.

64. Warner, *Portrait of Nelson,* 251.

65. To Admiral Lord Gardner, Commander in Chief, Ireland, 19 April 1805, *DLLN,* 6:413.

66. To William Beckford, 31 August 1805, *DLLN,* 7:22.

67. *DLLN,* 7:76.

68. Howarth and Howarth, *Nelson,* 203.

69. To Alexander Davison, 23 August 1799, *DLLN,* 7:cxci.

70. Clausewitz, *Vom Kriege,* 225; ibid., 214.

71. Ibid., 47; ibid.

72. Russell, *Nelson and the Hamiltons,* 312.

73. To Alexander Davison, 28 March 1804, *DLLN,* 5:476.

74. To Captain Cracraft, HM Ship *Anson,* 4 October 1804, *DLLN,* 6:217. See also *DLLN,* 6:129, 247.

75. To Lt. Harding Shaw, Commanding His Majesty's Brig *Spider,* 28 July 1804, *DLLN,* 6:118. See also Nelson to Capt. Ross Donnelly, HM Ship *Narcissus,* 14 July 1804, *DLLN,* 6:109–10.

76. *DLLN,* 6:271–72.

77. *DLLN,* 6:489.

78. *DLLN,* 7:148.

79. Letter to the Right Honorable George Rose, 6 October 1805, *DLLN,* 7:80.

80. *DLLN,* 7:148, 251.

81. Cf. Callo, *Legacy of Leadership,* 87, 88.

82. To Sir John Acton, 26 January 1805, *MHNP,* 2:253; Nelson to Captain Foote, HM Ship *Seahorse,* 8 June 1799, *DLLN,* 3:377.

83. Schom, *Trafalgar,* 354–55. Figures vary between sources. A recently published letter by Pvt. James Bagley, a Royal Marine on Nelson's flagship, dated 5 December 1805, reports: "we had 125 killed and wounded [on the flagship] and 1500 in the English fleet killed and wounded, and the enemy 12,000" (*ND* 7, pt. 4 [October 2000]: 238). John Keegan has slightly different totals in *Price of Admiralty,* 85–90. See also Terraine, *Trafalgar,* 162.

84. These figures have been calculated by this author during research for his many publications on Wehrmacht operational history and will appear as part of a forthcoming study on World War II attrition rates.

85. Rothenberg. *Art of War,* 81.

86. Patton, *War as I Knew It,* 335.

Chapter 5. Nelson and War on Land

1. Herold, *Mind of Napoleon,* 216.

2. "Sketch of My Life," 15 October 1799, *DLLN,* 1:3–15.

3. To Captain Locker, 12 August 1779, *DLLN,* 1:31.

4. "Sketch of My Life," 15 October 1799, *DLLN,* 1:9.

5. To William Locker, 12 August 1779, *DLLN,* 1:32.

6. Pocock, *Young Nelson,* 59.

7. Oman, *Nelson,* 30.

8. To Captain Locker, 23 January 1780, *DLLN,* 1:33.

9. Britton, "Nelson and the River San Juan," 216.

10. Official dispatch, 30 April 1780, *DLLN*, 1:9.

11. Pocock, *Horatio Nelson*, 91.

12. "Sketch of My Life," 15 October 1799, *DLLN*, 1:10.

13. Pocock, *Young Nelson*, 105, 106. Note that in the late 1700s, the term *sea officer* refers to a person serving on board a ship; *naval officer* generally meant someone who held a desk job at the admiralty.

14. Ibid., 112.

15. Dr. Moseley as quoted in Clarke and M'Arthur, *Life of Nelson*, 43.

16. Bradford, *Nelson*, 52.

17. Official dispatch, 30 April 1780, *DLLN*, 1:9.

18. "Sketch of My Life," 15 October 1799, *DLLN*, 1:10.

19. Clarke and M'Arthur, *Life of Nelson*, 43.

20. *DSJM*, 1 January 1794, 1:42

21. Fortescue, pt. 1 of *History*, vol. 4:179.

22. Gregory, *Ungovernable Rock*, 52, 57.

23. Fortescue, pt. 1 of *History*, 4:182.

24. *DSJM*, 1 March 1794, 1:56.

25. Lord Hood to Captain Nelson, 12 February 1794, in Lloyd, *Naval Miscellany*, 4:366–67.

26. Brenton, *Naval History*, 1:304.

27. Lt. Thomas Bowen to Sir William Hamilton, 18 February 1794, *NCL*, 49–50.

28. *DSJM*, 1:54, 55, 62, 63, 64; Elliot, *Life and Letters*, 2:236.

29. *DSJM*, 1 March 1794, 1:57.

30. Fortescue, pt 1 of *History*, 4:184; Pocock, *Nelson and Corsica*, 9.

31. Lord Hood to Lt. Gen. David Dundas, 6 March 1794, in Laughton, *Naval Miscellany* 4:371.

32. Fortescue, pt. 1 of *History*, 4:185.

33. Gregory, *Ungovernable Rock*, 66.

34. Letters to Fanny, 28 February and 4 March 1794, *NLW*, 103, 105.

35. Lady Minto to Lady Malmesbury, 7 September 1800, in Elliot, *Life and Letters*, 3:150.

36. *DSJM*, 1:66; Nelson to William Suckling, 1 March 1794, *DLLN*, 1:363.

37. Cf. *ATP-35 (B): Tactical Doctrine*, 3–5.

38. Ibid.

39. Elliot, *Life and Letters*, 2:234.

40. *DSJM*, 13 March 1794, 1:68.

41. *DSJM*, 16 March 1794, 1:71–72

42. Ibid.

43. To the Reverend Dixon Hoste, 3 May 1794, *MHNP*, 1:190–91.

44. To William Suckling, 7 February 1795, *DLLN*, 2:6.

45. To Fanny, 31 January 1795, *NLW*, 195.

46. Even Tom Pocock and Christopher Hibbert, authors of the two most popular

recent Nelson biographies, and certainly not hagiographers, gloss over their hero's manipulative actions without noting the severity in terms of military responsibilities. Pocock, *Horatio Nelson,* 114; Hibbert, *Nelson,* 95.

47. To Fanny, 1 May–4 May 1794, *NLW,* 109.

48. Lord Hood to Lt. Gen. David Dundas, 6 March 1794, in Laughton, *Naval Miscellany,* 4:371.

49. Lord Hood to Lt. Gen. David Dundas, 7 March 1794, in Laughton, *Naval Miscellany,* 4:371.

50. Lt. Gen. David Dundas to Lord Hood, 8 March 1794, in Laughton, *Naval Miscellany,* 4:371.

51. Minutes of Admiralty Orders to Captain Home Popham, 8 May 1798, *NCL,* 296–97. See also Minutes of Instructions to Vice Adm. Andrew Mitchell, July 1799, *NCL,* 320–22.

52. *DSJM,* 13 March 1794, 1:69.

53. Lord Hood to Captain Nelson, 8 March 1794, in Laughton, *Naval Miscellany,* 4:372.

54. Fortescue, pt. 1 of *History,* 4:188.

55. Ibid., 189; Nelson to the Reverend Mr. Nelson, 26 March 1794, *DLLN,* 1:375.

56. Elliot, *Life and Letters,* 2:234; Gregory, *Ungovernable Rock,* 71.

57. Howarth and Howarth, *Nelson,* 118.

58. To Commodore Linzee, *DLLN,* 1:379.

59. To Fanny, 22 April 1794, *NLW,* 108.

60. The total for the siege came to 1,058 barrels of powder, 11,923 shot, and 7,373 shells. Journal C, 23 May 1794, *DLLN,* 1:398; Howarth and Howarth, *Nelson,* 119.

61. *DSJM,* 23 April 1794, 1:81.

62. *DSJM,* 3 May 1794, 1:84.

63. Letter to the Reverend Dixon Hoste, 3 May 1794, *MHNP,* 1:190–91.

64. To Fanny, 1–4 May 1794, *NLW,* 109.

65. *DSJM,* 15 May 1794, 1:91.

66. Fortescue, pt. 1 of *History,* 4:190.

67. Lord Hood to Captain Nelson, 22 May 1794, in Laughton, *Naval Miscellany,* 4:400.

68. To Fanny, 20 May 1794, *NLW,* 112.

69. Fortescue, pt. 1 of *History,* 4:191.

70. Pocock, *Nelson and Corsica,* 18.

71. Journal C, 18 June 1794, *DLLN,* 1:408.

72. Pocock, *Nelson and Corsica,* 18.

73. Fortescue, pt. 1 of *History,* 4:192.

74. *DSJM,* 21 June 1794, 1:104; Nelson to Admiral Lord Hood, 21 June 1794, *DLLN,* 1:412.

75. One authority on early amphibious warfare estimates that it ordinarily required seventy men to haul a 32-pound cannon and fifty-six to haul a 24-pounder, and these

figures refer to cannons being pulled along a flat and even European road in fine weather (Harding, *Amphibious Warfare,* 167).

76. Cf. To Admiral Lord Hood, 23 June 1794, *DLLN,* 1:413.

77. To Major General Stuart, 20 June 1794, *DLLN,* 1:411.

78. To Fanny, 27 June 1794, *NLW,* 115.

79. Journal C, 10 August 1794, *DLLN,* 1:472.

80. To Admiral Lord Hood, 5 July 1794, *DLLN,* 1:421.

81. Journal C, 8 July 1794, *DLLN,* 1:426.

82. Muir notes that Napoleonic-era army batteries were commonly operated by 100–150 men and six or so officers, so we should see Stuart's batteries as typical of the era. See Muir, *Tactics and Battle,* 31.

83. *DSJM,* 13 July 1794, 1:110; Nelson letters to Admiral Lord Hood, 12 and 13 July 1794, *DLLN,* 1:433, 434.

84. Fortescue, pt. 1 of *Warfare,* 4:194.

85. To Admiral Lord Hood, 20 July 1794, *DLLN,* 1:449–50.

86. To the Right Honorable Sir Gilbert Elliot, 30 July 1794, *DLLN,* 1:461.

87. *DSJM,* 30 July 1794, 1:114.

88. Sir Gilbert Elliot to His Grace the Duke of Portland, 28 August 1794, in Laughton, *Naval Miscellany,* 4:417–18.

89. *DSJM,* 15 August 1794, 1:117.

90. *DSJM,* 29 June and 15 July 1794, 1:105, 110; Lord Hood to Captain Nelson, 16 July 1794, in Laughton, *Naval Miscellany,* 4:406.

91. To Lord Hood, 18 July 1794, *DLLN,* 1:445; to Lord Hood, 20 July 1794, in Laughton, *Naval Miscellany,* 4:409.

92. Lord Hood to Captain Nelson, 18 July 1794, in Laughton, *Naval Miscellany,* 4:407.

93. Letters to Lord Hood, 30 and 31 July 1794, *DLLN,* 1:460, 462; Gregory, *Ungovernable Rock,* 77–78.

94. Gregory, *Ungovernable Rock,* 77; Fortescue, part 1 of *History,* 4:195.

95. Lord Hood to General Stuart, 19 August 1794, in Laughton, *Naval Miscellany,* 4:416.

96. *DSJM,* 20 August 1794, 1:118.

97. To Fanny, 31 January 1795, *NLW,* 194.

98. To William Suckling, 7 February 1795, *DLLN,* 2:6.

99. Fortescue, part 1 of *History,* 4:197.

100. Cf. To the Duke of Clarence, 27 October 1795, *NLW,* 239–40.

101. To Fanny, 15 September 1795, *NLW,* 222.

102. To His Excellency Francis Drake, 11 April 1796, *DLLN,* 7:xlvi–xlvii.

103. White, "Nelson Ashore," 57. This comment is not to disparage White's assessment. It and White's other works remain at the forefront of Nelson scholarship.

104. To Adm. Sir John Jervis, 12 April 1797, *DLLN,* 2:378–81.

105. To Adm. Sir John Jervis, 12 April 1797, *DLLN,* 2:380.

106. White, *1797*, 100; Fortescue, part 2 of *History*, 4:797; Nelson to Adm. Sir John Jervis, 12 April 1797, *DLLN*, 2:379–80.

107. Miller records there being 740 seamen among the total landed force of about nine hundred. See Buckland, *Miller Papers*, 16, 19.

108. To Adm. Sir John Jervis, 7 June 1797, *DLLN*, 7:cxxxix–cxl.

109. See Houlding's excellent chapter on the training of marines, in his *Fit for Service*, 378–87.

110. And not many seamen wanted to learn the skills of soldiering. See Field. *Britain's Sea Soldiers*, 2:277.

111. Buckland, *Miller Papers*, 16.

112. For the development of marine artillery, see Field, *Britain's Sea Soldiers*, 2:259–75.

113. Buckland, *Miller Papers*, 15, 16, 19.

114. *DLLN*, 2:413.

115. Ireland, *Naval Warfare*, 132.

116. Gardiner, *Fleet Blockade*, 64.

117. To Thomas Troubridge, 20 July 1797, *DLLN*, 2:416–17.

118. Guimerá, *Nelson and Tenerif*, 7.

119. White, *1797*, 110–12.

120. Guimerá, *Nelson and Tenerif*, 7.

121. To Fanny, 4 August 1794, appended to that of 1 August, *NLW*, 118.

122. To His Excellency the Honorable John Trevor, 29 April 1796, *DLLN*, 7:lxi.

123. Field, *Britain's Sea Soldiers*, 2:262–63; Nelson to the Earl of St. Vincent, 25 May 1804, *DLLN*, 6:33–34.

124. To His Excellency the Honorable William Wyndham, 13 December 1799, *DLLN*, 4:136.

125. Glover, *Wellington*, 11.

Chapter 6. Coalition Warfare

1. A fine introduction to this topic is Neilson and Prete, *Coalition Warfare*.

2. Forester, *Lord Nelson*, 192–93.

3. Rodger, *War of the Second Coalition*, 66.

4. Ibid.; William Cathcart to Lord and Lady Cathcart, 27 August 1798, in Laughton, *Naval Miscellany*, 1:274.

5. Marquis de Niza to Sir William Hamilton, 11 August 1798, *MHNP*, 2:17.

6. To His Excellency the Marquis de Niza, 8 September 1798, *DLLN*, 3:118.

7. Marquis de Niza to Sir William Hamilton, 23 September 1798, *MHNP*, 2:20–21.

8. Earl of St. Vincent to Nelson, 2 July 1798, *DLLN*, 3:103–4.

9. Scott, *Recollections*, 119.

10. To Fanny, 23 June 1793, *NLW*, 83.

11. To Adm. Sir John Jervis, 9 June 1797, *DLLN*, 2:394.

12. Sir William Hamilton to Admiral the Earl of St. Vincent, 12 August 1798, *NLW,* 413–16.

13. To His Excellency the Marquis de Niza, 8 September 1798, *DLLN,* 3:118.

14. To the Right Honorable Sir William Hamilton, 7 September 1798, *MHNP,* 2:18.

15. To Admiral the Earl of St. Vincent, 1 September 1798, *DLLN,* 7:clxii.

16. *Office of the Chairman, Joint Chiefs of Staff, Department of Defense, Joint Publication 3-0, Doctrine for Joint Operations,* 1995, 6:7.

17. Earl of St. Vincent to Nelson, 2 July 1798, *DLLN,* 3:103–4.

18. Sabrosky, *Alliances,* 8.

19. Blanning, *French Revolutionary Wars,* 178–79.

20. Saul, *Russia and the Mediterranean,* 62.

21. Ibid., 60.

22. Ibid., 65; Rodger, *War of the Second Coalition,* 69.

23. McKnight, "Ushakov and the Ionian Republic," 9, 10, 38.

24. Ibid., 38.

25. Gardiner, *Nelson against Napoleon,* 58; Bennett, "Admiral Ushakov," 725; Rodger, *War of the Second Coalition,* 89; Saul, *Russia and the Mediterranean,* 79.

26. To His Excellency the Marquis de Niza, 13 September 1798, *DLLN,* 3:121.

27. To Earl Spencer, 16 September 1798, *DLLN,* 3:126.

28. To Earl Spencer, 25 and 29 September 1798, *DLLN,* 3:130, 137.

29. To His Excellency the Marquis de Niza, 4 October 1798, *DLLN,* 3:142.

30. To Admiral the Earl of St. Vincent, 30 September 1798, *DLLN,* 3:138.

31. To Lord Spencer, 9 October 1798, *NCL,* 116.

32. Knight, *Autobiography,* 1:129.

33. To His Excellency the Marquis de Niza, 24 October 1798, *MHNP,* 2:23.

34. To Admiral the Earl of St. Vincent, 25 January 1799, *DLLN,* 3:239.

35. To Fanny, 23 June 1793, *NLW,* 83.

36. To HRH The Duke of Clarence, 14 July–4 August 1793, *DLLN,* 1:312.

37. *DLLN,* 3:153.

38. To Admiral the Earl of St. Vincent, 3 November 1798, *DLLN,* 7:clxviii.

39. Ibid.

40. To Earl Spencer, 29 November 1798, *DLLN,* 3:180.

41. Ibid. The commodores (Mitchell, Stone, Campbell) were former British seamen who realized, correctly, that they were able to advance more quickly in foreign navies than in their own. Cf. Knight, *Autobiography,* 1:127; *DLLN,* 3:221.

42. Two Letters to His Excellency the Marquis de Niza, 24 October 1798, *DLLN,* 3:153–54.

43. Admiralty Minutes of Instructions to Lord St. Vincent, October 1798, *NCL,* 306–7; *DLLN,* 3:143–44.

44. Admiralty Minutes of Instructions to Lord St. Vincent, October 1798, *NCL,* 306–7.

45. Letters to John Spencer Smith, Esq., Constantinople, 7 and 26 October 1798, *DLLN,* 3:145–46, 159–60.

46. Sokol, "Nelson and the Russian Navy," 130.

47. To John Spencer Smith, Esq., Constantinople, 26 October 1798, *DLLN,* 3:159; Nelson to Francis Werry, Esq., Consul at Smyrna, 26 October 1798, *DLLN,* 3:160–61; letters to Admiral the Earl of St. Vincent, 4 October and 2 November 1798, *DLLN,* 7:clxvi and 3:165.

48. To Lord Spencer, 9 October 1798, *NCL,* 116.

49. To Admiral the Earl of St. Vincent, 27 September 1798, *DLLN,* 3:134; Nelson to the Inhabitants of Corfu, Cephalonia, Zante, and Cerigo, 9 October 1798, *DLLN,* 3:147–48.

50. To Captain Troubridge, 11 November 1798, *DLLN,* 3:169.

51. Ibid.

52. To Earl Spencer, 13 November 1798, *DLLN,* 3:172.

53. To His Excellency the Admiral Commanding the Ottoman Fleet, 17 November 1798, *DLLN,* 3:173.

54. To His Excellency the Admiral Commanding the Russian Fleet, 17 November 1798, 3:173.

55. McKnight, "Ushakov and the Ionian Republic," 82.

56. Ibid., 113.

57. Ibid., 120.

58. *DLLN,* 3:172, 174–75, 179–80.

59. Knight, *Autobiography,* 1:117.

60. To His Excellency Theodore Ouschakoff [Fedor Fedorovitch Ushakov], Vice Admiral of the Russian Squadron, 12 December 1798, *DLLN,* 3:197–98.

61. To John Spencer Smith, Esq., Constantinople, 17 December 1798, *DLLN,* 3:204.

62. Sokol, "Nelson and the Russian Navy," 131.

63. To Admiral the Earl of St. Vincent, 6 December 1798, *DLLN,* 3:184.

64. To Admiral the Earl of St. Vincent, 4 October 1798, *DLLN,* 7:clxvi.

65. To Simon Lucas, British Consul General at Tripoli, 20 March 1799, *NCL,* 120.

66. Ibid.

67. To the Grand Signior, 16 December 1798, *DLLN,* 3:202.

68. Ibid., 203.

69. To Vice Admiral Kadir Bey, 17 December 1798, *DLLN,* 3:203–4.

70. Ibid., 204.

71. To Lady Parker, 1 February 1799, *DLLN,* 3:248.

72. Cf. To Admiral the Earl of St. Vincent, 2 March 1799, *DLLN,* 3:275.

73. To Admiral the Earl of St. Vincent, 6 December 1798, *DLLN,* 3:183.

74. Knight, *Autobiography,* 125.

75. Ibid.

76. Ibid., 127–28.

77. To Rear Admiral His Excellency the Marquis de Niza, 21 December 1798, *DLLN,* 3:207.

78. To Captain Hope, His Majesty's Ship *Alcmene,* 21 December 1798, *DLLN,* 3:207.

79. To Rear Admiral His Excellency the Marquis de Niza, 22 December 1798, *DLLN,* 3:208.

80. To Admiral the Earl of St. Vincent, 31 December 1798, *NCL,* 117.

81. To the Right Honorable Earl Spencer, 1 January 1799, *DLLN,* 3:218.

82. *DLLN,* 3:215.

83. To the Right Honorable Earl Spencer, 1 January 1799, *DLLN,* 3:218.

84. To Commodore Duckworth, 7 January 1799, *DLLN,* 3:227.

85. That is clearly what he intended to convey in his letter to Lieutenant General Stuart, dated 7 January 1799 (*DLLN,* 3:228).

86. To the Marquis de Niza, 11 January 1799, *DLLN,* 3:231.

87. To the Marquis de Niza, 14 January 1799, *DLLN,* 3:233.

88. To Sir John Acton, 15 January 1799, *DLLN,* 3:233.

89. To the Marquis de Niza, 27 February 1799, *DLLN,* 3:271.

90. To Admiral the Earl of St. Vincent, 25 January 1799, *DLLN,* 3:239.

91. To Lt. Gen. Charles Stuart, 16 February 1799, *DLLN,* 3:267.

92. To His Excellency Rear Admiral the Marquis de Niza, 11 February 1799, *DLLN,* 3:262.

93. To His Excellency Rear Admiral the Marquis de Niza, 25 August 1799, *DLLN,* 3:464.

94. To His Excellency Don Rodrigo de Souza Coutinho, 27 August 1799, *DLLN,* 3:464–65.

95. To Captain Ball, Chief of the Island of Malta, 24 November 1799, *MHNP,* 2:77–78.

96. Disposition of the Squadron under the Command of Rear Admiral Lord Nelson, 11 September 1799, *DLLN,* 4:9–10.

97. To Captain Ball, 21 August 1799, *DLLN,* 3:456.

98. Cf. letters to Captain Ball, 4 and 9 February 1799, *DLLN,* 3:255, 260.

99. To His Highness the Bey of Tunis, 15 March 1799, *DLLN,* 3:293.

100. To the Marquis de Niza, 9 January 1799, *DLLN,* 3:231.

101. To Admiral the Earl of St. Vincent, 2 February 1799, *DLLN,* 3:251–52.

102. To the Marquis de Niza, 3 October 1799, *DLLN,* 4:39.

103. To Commo. Thomas Troubridge, 17 October 1799, *DLLN,* 4:58.

104. To His Excellency Rear Admiral the Marquis de Niza, 24 October 1799, *DLLN,* 4:62.

105. To His Excellency Rear Admiral the Marquis de Niza, 27 October 1799, *DLLN,* 4:71.

106. To His Excellency Don Rodrigo de Souza Coutinho, 2 January 1800, *DLLN,* 4:165.

107. To Commodore Duckworth, Port Mahon, 5 March 1799, *DLLN,* 3:280; McKnight, "Ushakov and the Ionian Republic," 132.

108. Saul, *Russia and the Mediterranean,* 109.

109. To Captain Ball, 21 January 1799, *DLLN,* 3:236.

110. To Captain Ball, 4 February 1799, *DLLN,* 3:255–56.

111. To His Excellency Theodore Ouschakoff [Ushakov], Vice Admiral of the Russian Fleet, 23 March 1799, *DLLN,* 3:304.

112. To Sir John Acton, 7 April 1799, *NNJ,* 11.

113. To Admiral the Earl of St. Vincent, 25 January 1799, *DLLN,* 3:238.

114. Rodger, *War of the Second Coalition,* 83.

115. Cf. To Captain Ball, 16 September 1799, *DLLN,* 4:19.

116. Bennett, "Admiral Ushakov," 727.

117. Ibid.

118. To His Excellency Theodore Ouschakoff, Vice Admiral of the Russian Fleet, 18 August 1799, *DLLN,* 3:448.

119. Saul, *Russia and the Mediterranean,* 117.

120. To Lord William Bentinck, 16 August 1799, *DLLN,* 3:446.

121. Ibid.

122. To Field Marshal Suwarrow [*sic*], 16 August 1799, *DLLN,* 3:448.

123. To Captain Ball, 5 September 1799, *DLLN,* 4:4.

124. Spencer to Nelson, 18 August 1799, *NLW,* 512.

125. To Earl Spencer, 6 September 1799, *DLLN,* Vol. 4:6.

126. Saul, *Russia and the Mediterranean,* 119.

127. Letter to J. Spencer Smith, 10 September 1799, *DLLN,* 4:7.

128. To His Imperial Majesty the Grand Signior, 10 September 1799, *DLLN,* 4:8.

129. To the Right Honorable the Earl of Elgin, 29 October 1799, *DLLN,* 4:76.

130. To Vice Admiral Ouschakoff, 1 November 1799, *DLLN,* 4:83.

131. Saul, *Russia and the Mediterranean,* 124.

132. Ibid., 125.

133. Ibid., 129.

134. Bennett, "Admiral Ushakov," 731.

135. Ibid.

Conclusion

1. The text of Secretary Dalton's inspiring speech is available from the United States Navy's Office of Information, and can be read online: www.chinfo.navy.mil/navpalib/people/secnav/dalton/speeches/traf1015.txt.

2. Ibid.

3. Trafalgar Night speech, Washington Navy Yard, 15 October 1997: www.chinfo.navy.mil/navpalib/people/flags/johnson_j/speeches/traf1015.txt.

4. Pocock, *Horatio Nelson,* 342.

Glossary

alliance A cooperative arrangement between two or more nations based on formal agreements, usually a treaty, to achieve broad, long-term foreign policy goals. An alliance is ordinarily more permanent than a coalition, which will usually dissolve or change after the attainment of some members' particular objectives.

blockade (1) The closing of enemy ports by naval vessels to cause economic disruption through the prevention of imports and exports. (2) The bottling up of enemy naval squadrons in ports in order to prevent them escaping and causing mischief or seeking battle elsewhere. In Nelson's time, *open blockade* involved keeping one's ships out to sea in the hope of intercepting slow moving merchant convoys and tempting enemy warships out of port to be ambushed. The far more arduous *close blockade* involved keeping ships constantly within the enemy port's visibility in order to discourage all attempts to move merchant or naval shipping in or out.

coalition An ad hoc cooperative arrangement between two or more nations to wage war together in order to achieve specific, usually short-term foreign policy goals. A coalition should not be confused with an *alliance.*

combined Denotes military or naval activities and operations that involve force elements from more than one nation. Combined operations are usually found in coalition or alliance contexts.

combined arms Denotes the coordinated employment of two or more branches or corps within an armed service (on land, it could be cavalry, infantry, and artillery; at sea it could be the coordinated use of ships of the line, bomb vessels, and frigates).

doctrine Most analysts agree that doctrine is a set of fundamental guiding ideas and principles, commonly written down to facilitate their ease of transmission and comprehension. The purpose of doctrine is to establish a common approach to various collective activities. In other words, doctrine provides a common reference point, language, and basic method, uniting the actions of many diverse elements into a team effort.

In the age of sail, the nearest thing to doctrine took the form of the British *Fighting Instructions,* which were far more prescriptive than any doctrine statements are today. First introduced in 1653, the earliest edition of the *Fighting Instructions* sought to impose a degree of order and consistency on the nature of battle by stressing the importance of the line-ahead formation. Successive editions introduced modifications when experiences of war revealed the need for guidelines on specific wind-based maneuvers. Yet the basic doctrine implicitly and explicitly proffered—that the line of battle and an established series of flag signals were the best means of reducing battle disorder, attaining concentration, and overcoming command and control deficiencies—changed little and, in fact, became rigidly formalized well before Nelson's time.

escalade An attack up-and-over fortress walls using scaling ladders, grapnels, and other such equipment. As a verb, escalade means to gain access to a fortified place by this type of attack.

firepower Firepower is the qualitative measure of the destructive power of weapons available to military or naval units. Counting the weapons alone will not allow firepower to be calculated. Calibers, velocities, and types must also be considered. For example, a mere fifteen 26-pounder cannons have far greater firepower than fifty 4-pounders.

Gallic, anti-Gallic Gallic (as in *Gaul*-ic) denotes the French people, nation, society, and culture. Anti-Gallic denotes dislike of the French and their ways.

Jacobin A contemporary term first used to describe the most radical French revolutionary grouping of the 1790–94 period, and later used (often less correctly) to describe French-sponsored or -inspired radical revolutionaries elsewhere in Europe.

joint Joint denotes activities or operations involving the closely synchronized employment of two or more service branches. In Nelson's time joint naturally meant the navy and the army, although not necessarily under a single commander. Nowadays, joint forces function under a single commander who has legally vested authority over all force elements in the theater. Not to be confused with *combined*.

leadership Denotes the art and activity of influencing others to do certain things. Most writers on leadership discuss two main types: *task oriented* and *people oriented*. (1) A task-oriented leader tends to organize his and the subordinates' roles with the accomplishment of the task as the highest priority. (2) A people-oriented leader makes his team members' needs the highest priority, even if this detracts somewhat from the task's completion. These behaviors are not mutually exclusive, and the best leadership type is one—as possessed by Nelson and others—that places high emphasis on both simultaneously.

marines The sea soldiers carried on board naval vessels and utilized by all contemporary European navies. Marines served as ships' policemen and conducted landings and cutting-out expeditions. During battle they provided random and volley fire with muskets and assisted in the operation of the ships' guns. Marines seldom took part in the day-to-day running of the ships and were never required to work aloft (except as sharpshooters).

military Persons or things pertaining to land forces, particularly armies and their functions. Differs from *navy, naval,* or *nautical,* which denote things pertaining to sea forces.

reconnaissance The act of obtaining as much information as possible about the position, direction, strength, composition, activities, and resources of the enemy's force. In the age of sail, frigates—single-gun-deck warships that functioned as scouts and convoy escort vessels due to their speed and flexibility—bore the lion's share of all naval reconnaissance responsibilities. Nimble yet well armed for their size, frigates constantly patrolled the ocean's expanses in search of enemy naval movements. They then reported their information with maximum haste to their parent squadron or fleet.

ships of the line Also known as *line of battle* ships and later simply called battleships, these were naval vessels deemed large and powerful enough to sail in a line of battle (usually with sixty-four to one hundred or more guns, seventy-four being the norm). For more than a century before Nelson, fleet tactics had been based on the line of battle. This involved fleets or squadrons approaching in orderly lines, sometimes stretching out for several miles or more. They engaged each other obliquely (from a slight and ever-closing angle) or in parallel, with each ship seeking to place itself alongside a similar or weaker opponent. Firepower dominated this linear form of battle, and victory almost always went to the side that possessed the most and fastest cannons, or "guns" as seamen usually called them.

shot, shells (1) Shot, sometimes known as roundshot, were the most basic and common artillery ammunition. They were solid cast-iron metal balls designed simply to smash through whatever they hit. (2) Shells, on the other hand, were exploding ammunition. They had attached fuses, which ignited instantly when the guns fired. The gunpowder-filled shells exploded after only a few seconds, which made them dangerous in close-range artillery roles.

Given that most naval battles occurred at very close range, and using shells was risky amid the fiery chaos of battle at sea, few ships ever fired shells. Bomb vessels used them, as did the ships' guns that were sometimes taken ashore and employed in place of field artillery.

tactics Tactics are the plans made, directions given, and actions taken before and during specific engagements or battles. The aim of tactics is to ensure that the positions, weapons, and actions of all personnel and units within a combat force are coordinated with each other in order to achieve maximum impact on the enemy. The *Fighting Instructions* served as the basic reference for naval tactics in battles, and departure from them was seldom attempted, approved, or left unpunished (unless resulting in a decisive victory).

Bibliography

Allen, William, and Edmund McClure. *Two Hundred Years: The Society for Promoting Christian Knowledge 1698–1898*. New York: Burt Franklin, 1898.

ATP-35: [NATO] Land Force Tactical Doctrine. N.p., n.d.

Bennett, Geoffrey. "Admiral Ushakov: Nelson's Russian Ally." *History Today* 21, no. 10 (October 1971): 724–31.

———. *Nelson the Commander*. New York: Scribner's, 1972.

Berckman, Evelyn. *Nelson's Dear Lord: A Portrait of St. Vincent*. London: Macmillan, 1962.

Biancareddu, Don Antonio. Letter to Nelson. 22 October 1804. In *Trafalgar Chronicle: The Yearbook of the 1805 Club* 11 (2001): 36.

Blanning, Tim C. W. *The French Revolutionary Wars, 1787–1802*. London: Arnold, 1996.

Bonner-Smith, David, ed. *Letters of Admiral of the Fleet the Earl of St. Vincent whilst First Lord of the Admiralty, 1801–1804*. Vol. 55. London: Navy Records Society, 1922.

———, ed. *Letters of Admiral of the Fleet the Earl of St. Vincent whilst First Lord of the Admiralty, 1801–1804*. Vol. 61. London: Navy Records Society, 1926.

Bradford, Ernle. *Nelson: The Essential Hero*. London: Macmillan, 1977.

Brenton, Edward. *The Naval History of Great Britain*. London: Henry Colburn, 1837.

Britton, C. J. "Nelson and the River San Juan." *The Mariner's Mirror* 28 (July 1942): 213–21.

Broadley, Alexander M., and R. G. Bartelot. *Nelson's Hardy: His Life, Letters and Friends*. London: John Murray, 1909.

Buckland, Kirstie, ed. *The Miller Papers*. Shelton: The 1805 Club, 1999.

Callo, Joseph. *Legacy of Leadership: Lessons from Admiral Lord Nelson*. Central Point, Ore.: Hellgate, 1999.

———. *Nelson Speaks: Admiral Lord Nelson in His Own Words*. Annapolis: Naval Institute Press, 2001.

Clarke, James S., and John M'Arthur. *The Life of Admiral Lord Nelson, K.B., from His Lordship's Manuscripts*. London: Cadell and Davies, 1810.

Clausewitz, Carl von. *Vom Kriege: Hinterlassenes Werk* (On War: Posthumous Work). Ungekürzter text (Unexpurgated [and complete] text). 1832. Reprint, Berlin: Ullstein Verlag, 1999.

Coleman, Terry. *Nelson: The Man and the Legend*. London: Bloomsbury, 2001.

Connelly, Owen. *Blundering to Glory: Napoleon's Military Campaigns*. Rev. ed. Wilmington, Del.: Scholarly Resources, 1999.

Dann, John, ed. *The Nagle Journal: A Diary of the Life of Jacob Nagle, Sailor, from the Year 1775 to 1841*. New York: Weidenfeld and Nicolson, 1988.

Dawson, Warren, ed. *The Nelson Collection at Lloyds: A Description of the Nelson Relics and a Transcript of the Autograph Letters and Documents of Nelson and His Circle and of Other Naval Papers of Nelson's Period*. London: Macmillan for Lloyds, 1932.

Dickinson, Harry Thomas, ed. *Britain and the French Revolution, 1789–1815*. London: Macmillan, 1989.

Dixon, Norman. *On the Psychology of Military Incompetence*. New York: Basic, 1976.

Dugan, James. *The Great Mutiny*. New York: Putnams, 1965.

Durant, H. "Lord Nelson's Visit to Monmouth." *Memorials of Monmouth*. Vol. 4. Ross-on-Wye: Monmouth and District Field Club and Antiquarian Society, n.d.

Elliot, Gilbert. *Life and Letters of Sir Gilbert Elliot, First Earl of Minto from 1751 to 1806, Edited by His Great-Niece the Countess of Minto in Three Volumes*. London: Longmans, Green, 1874.

Evans, Michael, and Alan Ryan, eds. *The Human Face of Warfare: Killing, Fear and Chaos in Battle*. St. Leonards, England: Allen and Unwin, 2000.

"Extract of a Letter." *Edinburgh Advertiser*, 24 April 1801, 261.

Field, Cyril. *Britain's Sea Soldiers: A History of Royal Marines and Their Predecessors and of Their Services in Action, Ashore and Afloat*. Liverpool: Lyceum, 1924.

Finlayson, Cdr. John. "A Signal Midshipman at Copenhagen." *Trafalgar Chronicle: The Yearbook of the 1805 Club*, ed. Colin White, 11 (2001): 80–97.

Fitchett, William H. *Nelson and His Captains*. London: Smith, Elder, 1906.

Foreman, Laura, and Ellen B. Phillips. *Napoleon's Lost Fleet: Bonaparte, Nelson and the Battle of the Nile*. London: Discovery Books, 1999.

Forester, C. S. *Lord Nelson*. Indianapolis: Bobbs-Merrill, 1929.

Fortescue, John Willam. *A History of the British Army*. Vol. 4, 2 pts. London: Macmillan, 1915.

Fremantle, Anne, ed. *The Wynne Diaries, 1789–1820*. 3 vols. London: Oxford University Press, 1936, 1937, 1940.

Gardiner, Robert, ed. *Fleet Blockade and Battle: The French Revolutionary War, 1793–1797*. Annapolis: Naval Institute Press, 1996.

———, ed. *Nelson against Napoleon: From the Nile to Copenhagen, 1798–1801*. London: Chatham, 1997.

Gash, Norman. "The Fourth Wellington Lecture." *Wellington Anecdotes: A Critical Survey*. Southampton, England: University of Southampton, 1992.

Glover, Michael. *Wellington as Military Commander*. London: Penguin, 2001.

The Great Commanders, "Nelson." Great Britain: Channel 4 television, 1993.

Gregory, Desmond. *The Ungovernable Rock: A History of the Anglo-Corsican Kingdom and Its Role in Britain's Mediterranean Strategy during the Revolutionary War, 1793–1797*. Cranbury: Associated University Presses, 1985.

Guimerá, Augustín. *Nelson and Tenerife, 1797*. Shelton: The 1805 Club, 1999.

Gutteridge, Harold, ed. *Nelson and the Neapolitan Jacobins: Documents Relating to the Suppression of the Jacobin Revolution at Naples, June 1799*. Vol. 25. Navy Records Society, 1903.

Gwyther, J. "Nelson's Gift to La Maddalena." *Trafalgar Chronicle: The Yearbook of the 1805 Club* 10 (2000): 47–56.

———. " . . . in order to thank your Excellency. . . ." *Trafalgar Chronicle: The Yearbook of the 1805 Club* 11 (2001): 35–37.

Harding, Richard. *Amphibious Warfare in the Eighteenth Century*. Woodbridge: Boydell Press, A Royal Historical Society Publication, 1991.

Harper, Glyn, and Garth Pratten. *Still the Same: Reflections on Active Service from Bardia to Baidoa*. Georges Heights: Australian Army Doctrine Center, 1996.

Hayward, Joel. "Horatio Lord Nelson's Warfighting Style and the Maneuver Warfare Paradigm." *Defence Studies* 1, no. 2 (summer 2001): 15–37.

Herold, J. Christopher, ed. *The Mind of Napoleon: A Selection from His Written and Spoken Words*. New York: Columbia University Press, 1955.

Hibbert, C. *Nelson: A Personal History*. London: Viking, 1994.

———. *Wellington: A Personal History*. London: HarperCollins, 1997.

Hooker, Richard D. Jr., ed. *Maneuver Warfare: An Anthology*. Novato: Presidio Press, 1993.

Hore, Peter, ed. *Seapower Ashore: 200 Years of Royal Navy Operations on Land*. London: Chatham, 2001.

Houlding, J. A. *Fit for Service: The Training of the British Army, 1715–1795*. Oxford: Oxford University Press, 1981.

Howarth, David, and Stephen Howarth. *Nelson: The Immortal Memory*. London: Dent, 1988. Reprint, London: Conway Classics, 1997.

Hughes, Edward, ed. *The Private Correspondence of Admiral Lord Collingwood*. London: Navy Records Society, 1957.

Ireland, Bernard. *Naval Warfare in the Age of Sail: War at Sea, 1756–1815*. New York: Norton, 2000.

Jennings, Louis John, ed. *The Croker Papers: The Correspondence and Diaries of the Late Right Honourable J. W. Croker, LL.D., F.R.S., Secretary to the Admiralty from 1809 to 1830*. London: John Murray, 1884.

Jordan, G. "Admiral Nelson as Popular Hero: The Nation and the Navy, 1795–1805." *New Aspects of Naval History: Selected Papers from the 5th Naval History Symposium*, pp. 109–19. Baltimore: Nautical and Aviation, 1985.

Keegan, John. *The Mask of Command*. New York: Viking-Penguin, 1987.

———. *The Price of Admiralty: War at Sea from Man of War to Submarine*. London: Hutchinson, 1988.

Kennedy, Ludovic. *Nelson's Band of Brothers*. London: Odhams, 1951.

King, Dean, ed. *Every Man Will Do His Duty: An Anthology of Firsthand Accounts from the Age of Nelson*. New York: Henry Holt, 1997.

Kirby, Brian S. "Nelson and American Merchantmen in the West Indies, 1784–1787." *The Mariner's Mirror* 75, no. 2 (1989): 137–47.

Knight, Cornelia. *Autobiography of Miss Cornelia Knight, Lady Companion to the Princess Charlotte of Wales, with Extracts from Her Journals and Anecdote Books*. 2 vols. London: W. H. Allen, 1861.

Laughton, John Knox, ed. *The Naval Miscellany*. Vol. 1. London: Navy Records Society, 1902.

———, ed. *The Naval Miscellany*. Vol. 2. London: Navy Records Society, 1912.

Lavery, Brian. *Nelson's Navy: The Ships, Men, and Organization, 1793–1815*. Annapolis: Naval Institute Press, 1994.

———. *Nelson and the Nile: The Naval War against Bonaparte, 1798*. London: Chatham, 1998.

Leonhard, Robert. *The Art of Maneuver: Maneuver-Warfare Theory and AirLand Battle*. Novato: Presidio Press, 1991.

Lewis, Michael. *A Social History of the Navy, 1793–1815*. London: Allen and Unwin, 1960.

Liddell Hart, Basil H. *Thoughts on War*. London: Faber, 1943.

Lind, William S. *Maneuver Warfare Handbook*. Boulder, Colo.: Westview Press, 1985.

———. "Misconceptions of Maneuver Warfare." *Marine Corps Gazette* 72, no. 1 (January 1988): 17.

Lloyd, Christopher, ed. *The Naval Miscellany*. Vol. 4. London: Navy Records Society, 1952.

Maccunn, Francis John. *The Contemporary English View of Napoleon*. London: G. Bell, 1914.

McKnight, James Lawrence. "Admiral Ushakov and the Ionian Republic: The Genesis of Russia's First Balkan Satellite." Ph.D. diss., University of Wisconsin, 1965.

Macleod, Emma V. *A War of Ideas: British Attitudes to the Wars against Revolutionary France, 1792–1802*. Aldershot, England: Ashgate, 1998.

McLynn, Frank. *Napoleon: A Biography*. London: Pimlico, 1997.

Maffeo, Steven E. *Most Secret and Confidential: Intelligence in the Age of Nelson*. Annapolis: Naval Institute Press, 2000.

Mahan, Alfred Thayer. *The Life of Nelson, the Embodiment of the Sea Power of Great Britain*. London: Samson Low, Marston, 1899.

Matcham, Mary Eyre. *The Nelsons of Burnham Thorpe: A Record of a Norfolk Family Compiled from Unpublished Letters and Notebooks, 1787–1842*. London: John Lane, the Bodley Head, 1911.

Maurice, John Frederick, ed. *The Diary of Sir John Moore.* 2 vols. London: Edward Arnold, 1904.

MCDP 1: Warfighting. N.p.: United States Marine Corps, 1997.

MCDP 1-3: Tactics. N.p.: United States Marine Corps, 1997.

Moorhouse, Esther Hallam. *Nelson in England: A Domestic Chronicle.* London: Chapman and Hall, 1913.

Morrison, Alfred, ed. *The Collection of Autograph Letters and Historical Documents Formed by Alfred Morrison (Second Series, 1882–1893), The Hamilton and Nelson Papers.* Vol 1: *1756–1797;* Vol 2: *1798–1815.* Privately printed 1893, 1894.

Morriss, Roger. *Nelson: The Life and Letters of a Hero.* London: Collins and Brown, 1996.

Muir, Rory. *Tactics and the Experience of Battle in the Age of Napoleon.* New Haven: Yale University Press, 1998.

Naish, George, ed. *Nelson's Letters to His Wife and Other Documents 1785–1831.* London: Routledge and Kegan Paul in conjunction with the Navy Records Society, 1958.

Neilson, Keith, and Roy A. Prete. *Coalition Warfare: An Uneasy Accord.* Waterloo, Ont.: Wilfred Laurier University Press, 1983.

Nicolas, Nicholas Harris, ed. *The Dispatches and Letters of Vice Admiral Lord Viscount Nelson, with Notes by Sir Nicholas Harris Nicolas, GCMG.* 7 vols. 1844–1846. Reprint, London: Chatham, 1998.

Oman, Carola. *Nelson.* London: Hodder and Stoughton, 1947. Reprint, London: History Book Club, 1967.

Orde, Dennis A. *Nelson's Mediterranean Command.* Durham, England: Pentland Press, 1997.

Palmer, Michael A. "Sir John's Victory: The Battle of Cape St. Vincent Reconsidered." *The Mariner's Mirror* 77, no. 1 (February 1991): 31–46.

———. "The Soul's Right Hand: Command and Control in the Age of Fighting Sail, 1652–1827." *The Journal of Military History* 61 (October 1997): 679–706.

Parsons, George Samuel. *Nelsonian Reminiscences: A Dramatic Eye-witness Account of the War at Sea, 1795–1810.* 1843. London: Chatham, 1998.

Patton, George S. *War as I Knew It.* Boston: Houghton Mifflin, 1947. Reprint, New York: Bantam, 1979.

Pocock, Tom. *Remember Nelson: The Life of Captain Sir William Hoste.* London: Collins, 1977.

———. *The Young Nelson in the Americas.* London: Collins, 1980.

———. *Horatio Nelson.* London: Pimlico, 1994.

———. *Nelson and the Campaign in Corsica.* Hoylake, England: The 1805 Club, 1994.

Pope, Dudley. *Life in Nelson's Navy.* Annapolis: Naval Institute Press, 1983.

Rawson, Geoffrey, ed. *Nelson's Letters.* London: Dent, 1960.

Rodger, Alexander Bankier. *The War of the Second Coalition, 1798–1801: A Strategic Commentary.* Oxford: Clarendon Press, 1964.

Rodger, N. A. M. *The Wooden World: An Anatomy of the Georgian Navy.* New York: Norton, 1986.

Rose, John Holland. *Lord Hood and the Defence of Toulon.* Cambridge: Cambridge University Press, 1922.

Rothenberg, Gunther E. *The Art of War in the Age of Napoleon.* Bloomington: Indiana University Press, 1980.

Russell, Jack. *Nelson and the Hamiltons.* Harmondsworth, England: Penguin, 1972.

Sabrosky, Alan Ned, ed. *Alliances in U.S. Foreign Policy: Issues in the Quest for Collective Defence.* London: Westview, 1988.

Saul, Norman E. *Russia and the Mediterranean, 1797–1807.* Chicago: University of Chicago Press, 1970.

Schneider, Barry R., and Laurence E. Grinter, eds. *Battlefield of the Future: 21st Century Warfare Issues.* Maxwell Air Force Base: Air War College, 1998.

Schom, Alan. *Trafalgar: Countdown to Battle, 1803–1805.* Oxford: Oxford University Press, 1990.

———. *Napoleon Bonaparte.* New York: HarperCollins, 1997.

Scott, A. J. *Recollections of the Life of the Rev. A. J. Scott, D.D., Lord Nelson's Chaplain.* London: Saunders and Otley, 1842.

Shannon, David. "Two Missing Nelson Letters Reconstructed." Part 9. *The Nelson Dispatch* 7 (January 2002): 595–98.

Sheffield, G. D., ed. *Leadership and Command: The Anglo-American Experience since 1861.* London: Brassey's, 1997.

Simpkin, Richard E. *Race to the Swift: Thoughts on Twenty-first Century Warfare.* London: Brassey's, 1984.

Sokol, A. E. "Nelson and the Russian Navy." *Military Affairs* 13, no. 3 (autumn 1949): 129–37.

Terraine, John. *Trafalgar.* London: Sidgwich and Jackson, 1976. Reprint, Ware, England: Wordsworth, 1998.

Thursfield, James Richard. *Nelson and Other Naval Studies.* London: John Murray, 1920.

Tracy, Nicholas. *Nelson's Battles: The Art of Victory in the Age of Sail.* London: Chatham, 1996.

———. *The Naval Chronicle: The Contemporary Record of the Royal Navy at War.* 5 vols. London: Chatham, 1998–1999.

Van Creveld, Martin. *Air Power and Maneuver Warfare.* Montgomery, Ala.: Air University Press, 1994.

Warner, Oliver. *A Portrait of Lord Nelson.* London: Chatto and Windus, 1959.

———. *Nelson's Last Diary and the Prayer before Trafalgar: A Facsimile*. Kent, Ohio: Kent State University Press, 1971.

Weller, Jack. *On Wellington: The Duke and His Art of War*. London: Greenhill, 1998.

White, Colin. *1797: Nelson's Year of Destiny*. Stroud: Sutton, in association with the Royal Naval Museum, 1998.

———. "Nelson and Shakespeare." Part 3. *The Nelson Dispatch* 7 (July 2000): 145–50.

———. "Nelson Ashore 1780–97." In *Seapower Ashore: 200 Years of Royal Navy Operations on Land*. Edited by Peter Hore. London: Chatham, 2001.

Winton, John, ed. *An Illustrated History of the Royal Navy*. London: Salamander, 2000.

Wybourn, T. M. *Sea Soldier: An Officer of Marines with Duncan, Nelson, Collingwood, and Cockburn: The Letters and Journals of Major T. Marmaduke Wybourn, RM, 1797–1813*. Tunbridge Wells, England: Parapress, 2000.

Index

About the Author

Until mid-2002, **Dr. Joel Hayward** was a senior lecturer in defense and strategic studies at Massey University in his native New Zealand. Currently, he works as a freelance analyst and writer and is a research associate with the United States Air Force. Dr. Hayward earned his master's degree and his doctorate at Canterbury University.

Dr. Hayward has written several books, including a highly praised analysis of the Luftwaffe's role in the Battle of Stalingrad, as well as dozens of peer-reviewed articles on such topics as military history, strategy, operational art, and command. Dr. Hayward also has a keen interest in Slavic history and culture. His work, which includes poetry and fiction, has been published in various languages, including German, Russian, and Serbo-Croatian.

Dr. Hayward lives with his wife, Kathy, and their daughters in Palmerston North, New Zealand.

The Naval Institute Press is the book-publishing arm of the U.S. Naval Institute, a private, nonprofit, membership society for sea service professionals and others who share an interest in naval and maritime affairs. Established in 1873 at the U.S. Naval Academy in Annapolis, Maryland, where its offices remain today, the Naval Institute has members worldwide.

Members of the Naval Institute support the education programs of the society and receive the influential monthly magazine *Proceedings* and discounts on fine nautical prints and on ship and aircraft photos. They also have access to the transcripts of the Institute's Oral History Program and get discounted admission to any of the Institute-sponsored seminars offered around the country.

The Naval Institute also publishes *Naval History* magazine. This colorful bimonthly is filled with entertaining and thought-provoking articles, first-person reminiscences, and dramatic art and photography. Members receive a discount on *Naval History* subscriptions.

The Naval Institute's book-publishing program, begun in 1898 with basic guides to naval practices, has broadened its scope to include books of more general interest. Now the Naval Institute Press publishes about one hundred titles each year, ranging from how-to books on boating and navigation to battle histories, biographies, ship and aircraft guides, and novels. Institute members receive significant discounts on the Press's more than eight hundred books in print.

Full-time students are eligible for special half-price membership rates. Life memberships are also available.

For a free catalog describing Naval Institute Press books currently available, and for further information about subscribing to *Naval History* magazine or about joining the U.S. Naval Institute, please write to:

Membership Department
U.S. Naval Institute
291 Wood Road
Annapolis, MD 21402-5034
Telephone: (800) 233-8764
Fax: (410) 269-7940
Web address: www.navalinstitute.org